STATE VARIABLES AND COMMUNICATION THEORY

STATE VARIABLES AND COMMUNICATION THEORY

ARTHUR B. BAGGEROER

Research Monograph No. 61
The M.I.T. Press
Cambridge, Massachusetts, and London, England

Set in Monotype Times Roman
Printed and bound in the United States of America by
The Colonial Press, Inc., Clinton, Massachusetts.

ISBN 0 262 02060 2 (hardcover)

Library of Congress catalog card number: 73-107986

Contents

Preface xi

1. Introduction 1

1.1 State Variables and Communication Theory 1
1.2 Organization 4
1.3 Notation 7

2. State Variable Random Processes 8

2.1 Generation of Random Processes with State Variables 8
2.2 Covariance Functions for State Variable Processes 10
2.3 The Derivation of the Differential Equations for the Covariance Operator 18

3. Homogeneous Fredholm Integral Equations 23

3.1 The State Variable Solution to Homogeneous Fredholm Integral Equations 24
3.2 The Fredholm Determinant Function 31
3.3 Examples 35
Example 3.1 The Wiener Process 35
Example 3.2a A First-Order Stationary Process: The First-Order Butterworth 36
Example 3.2b A First-Order Nonstationary Process 38
Example 3.3 A Second-Order Stationary Process 39
Example 3.4 The Higher-Order Butterworth Processes 43
3.4 Discussion of Results for Homogeneous Fredholm Integral Equations 54

4. Inhomogeneous Fredholm Integral Equations 56

4.1 Communication Models for Detecting Signals in Colored Noise 57
4.2 The State Variable Solution to Inhomogeneous Fredholm Integral Equations 61
4.3 Methods of Solving the Differential Equations for the Inhomogeneous Fredholm Equation 63
Method 1 64
Method 2 66
Method 3 68
Summary of Methods 70
4.4 Examples of Solutions to the Inhomogeneous Fredholm Equation 70
Example 4.1 $g(t)$ for a Wiener Process 70
Example 4.2 $g(t)$ for a First-Order Stationary Spectrum 72
Example 4.3 $g(t)$ for a Second-Order Stationary Process 75
4.5 Discussion of Results for Inhomogeneous Fredholm Integral Equations 78

5. Optimal Signal Design for Colored Noise Channels via State Variables 81

5.1 Problem Formulation 82
5.2 The Application of the Minimum Principle 85
5.3 Optimal Signal Design for Additive Signal-Independent Noise Channels 96
Example 5.1 Signal Design for Channels with a First-Order Noise Spectrum 102
Example 5.2 Signal Design with a Second-Order Spectrum 109
Signal Design with a Hard Bandwidth Constraint 114
5.4 Optimal Signal Design for Doppler Spread Environments 116
5.5 Summary and Discussion 118

6. Linear Smoothing and Filtering with Delay 121

6.1 The Optimal Linear Smoother 123
6.2 Covariance of Error for the Optimum Smoother 128
6.3 Examples of the Optimal Smoother Performance 134
Example 6.1 Covariance of Error for a Wiener Process 134
Example 6.2 Covariance of Error for a First-Order Process 135
Case a $P_0 = P$ (Stationary Process) 137
Case b $P_0 = 2P/\Lambda + 1$ (Steady-State Realizable Filtering Error) 137
Case c $P_0 = 0$ (Known Initial State) 140
Example 6.3 Covariance of Error for a Second-Order Process 140
6.4 Filtering with Delay 144
6.5 Performance of the Filter with Delay 149
6.6 Example of Performance of the Filter with Delay 153
6.7 Discussion and Conclusions for the Optimal Smoother and Filter with Delay 155

7. Smoothing and Filtering for Nonlinear Modulation System 156

7.1 State Variable Model of Linear Systems 157
7.2 Smoothing for Nonlinear Modulation Systems 161
7.3 An Approximate Solution to the Realizable Filter 163

Appendix A. Computation of the Exponential Matrix 170

Appendix B. Complex Random Process Generation 172

B.1 Random Process Generation with a Complex Notation 173
B.2 Stationary Random Processes 178
B.3 Summary 178
Example B.1 179
Example B.2 180

Bibliography 187

Index 193

Foreword

This is the sixty-first volume in the M.I.T. Research Monograph Series published by the M.I.T. Press. The objective of this series is to contribute to the professional literature a number of significant pieces of research, larger in scope than journal articles but normally less ambitious than finished books. We believe that such studies deserve a wider circulation than can be accomplished by informal channels, and we hope that this form of publication will make them readily accessible to research organizations, libraries, and independent workers.

Howard W. Johnson

Preface

It is the purpose of this monograph to demonstrate how state variable concepts can be successfully applied to a variety of problems in communication theory. Although these concepts have become firmly established in modern control theory, their application as discussed in the communication theory literature has not been nearly as extensive. State variable concepts offer at least two advantages to the communication theorist. First, they provide another view into a problem, often one which brings new insight. Second, they often lead to solution procedures that are readily adaptable to the digital computer. In many cases, the methods are not only straightforward to program, but they also often offer significant savings in the actual computation time required.

At first we focus our attention here on their utility in solving integral equations. These equations are of fundamental importance in communication theory; consequently, our concepts and results are useful in a large number of problems of interest. We then consider two specific applications. The first concerns optimal signal design for colored noise channels while the second considers linear estimation theory.

For background we assume that the reader is familiar with detection and estimation theory at a level as discussed by Van Trees (H. L. Van Trees, *Detection, Estimation, and Modulation Theory, Part I*, John Wiley & Sons, Inc., New York, 1968) and with deterministic state variable concepts as discussed by Zadeh and Desoer (L. A Zadeh and C. Desoer, *Linear System Theory*, McGraw-Hill Book Company, Inc., New York, 1966), by Athans and Falb (M. Athans and P. L. Falb,

Optimal Control, McGraw-Hill Book Company, Inc., New York, 1966), or by DeRusso, Roy, and Close (P. M. DeRusso, R. J. Roy, and C. M. Close, *State Variables for Engineers*, John Wiley & Sons, Inc., New York, 1965). We developed the needed state variable results as related to random processes. A more tutorial development can also be found in Van Trees. When we discuss the estimation problems of smoothing and filtering with delay, we also assume that the reader is familiar with Kalman and Bucy's results on optimal filtering (R. E. Kalman and R. Bucy, "New Results in Linear Filtering and Prediction Theory," *ASME J. Basic Eng.*, **83**, 95–108 (1961)).

This monograph is essentially a revised version of the author's doctoral thesis (A. B. Baggeroer, "State Variables, the Fredholm Theory, and Optimal Communication." Sc.D. Thesis, Department of Electrical Engineering, Massachusetts Institute of Technology, Cambridge, Mass., 1968). In this context it is a pleasure to acknowledge the patient supervision and interest of Professor Harry Van Trees. It is also a pleasure to acknowledge the interest of Professors Wilbur Davenport and William Siebert, who served as readers on my thesis committee.

Several people made valuable contributions to my thesis. I profited greatly from my association with my fellow student Lewis Collins and with Professor Donald Snyder. Discussions with Theodore Cruise have clarified several issues of the material.

Earlier drafts of the material were typed by Miss Camille Tortorici and Mrs. Vera Conwicke, while the final manuscript was done by Mrs. Enid Zollweg. Their efforts are sincerely appreciated.

The work was supported in part by the National Science Foundation, a U.S. Navy Purchasing Contract, and by the Joint Services Electronics Program. I am grateful for this support. The computations were performed at the M.I.T. Computation Center.

STATE VARIABLES AND COMMUNICATION THEORY

1 Introduction

1.1 State Variables and Communication Theory

A state variable description of systems and random processes offers several advantages from both theoretical and practical viewpoints. From a theoretical aspect, such a description provides a very general characterization in terms of which a large class of systems, possibly time varying and nonlinear, can be modeled. Many powerful and elegant statements can be made with regard to systems described in this manner. From a practical aspect, they often provide a more representative physical description of the actual dynamics of the systems involved. More importantly, a state variable approach often leads to solution techniques that are readily implemented on a computer. This is highly desirable when specific numerical results are required.

The essential feature of a state variable approach is that the systems and processes of interest are described in terms of differential equations and their excitation, which is usually a white noise process. This is in contrast to the impulse response and covariance function description of systems and processes commonly used in the analysis of communication problems. Since a computer is ideally suited for integrating differential equations, one can easily see how a state variable formulation leads to effective computational solution methods.

In the area of automatic control, state variable concepts have been used extensively, so much so that they are the approach used in the majority of problems now studied. In communication theory, by contrast, they are not employed nearly as extensively. While a state variable description is certainly not appropriate in many situations, there are a large number of problems in communication theory where these concepts can be used advantageously. This monograph is directed to those people in communication theory who want to exploit some of the concepts and methods of state variables in the analysis of their problems.

The use of state variables is not novel in that they have already provided effective solutions to several important problems in communication theory. Undoubtedly, the most significant of these is the original work of Kalman and Bucy on optimal linear minimum mean square error realizable filtering.[35] In the classical approach, as used by Wiener, the random processes are represented in terms of their covariances. The impulse response of the optimum filter then is determined in terms of these covariances, or their associated spectra. Consequently, the optimal estimate is the result of the explicit operation of this impulse response upon the observed signal. In contrast, Kalman and Bucy represented the random processes in terms of state variables. They then found a structure for the optimum filter in which the desired estimate is specified implicitly as the solution to a set of differential equations. The principal advantage here is that it is usually more convenient, especially computationally, to implement solving the differential equations than it is to realize the operation implied by the impulse response.

Starting with the concepts introduced in their papers, several people have used state variable techniques in analyzing problems concerned with the detection and estimation of random processes. Particularly noteworthy contributions have been made by Schweppe in the detection of Gaussian random signals in Gaussian noise,[52] Kushner in the general theory of nonlinear filtering,[40] and Snyder in the application of state variable, nonlinear filtering to communication systems.[61] Certainly, many other results published in the control literature are also relevant to communication problems.

Here, we are principally interested in how state variables can be used effectively to solve several of the integral equations that frequently appear in communications theory. These equations and their associated theory assume an important role in communications. One often encounters situations where a fundamental result of a particular analysis is succinctly stated, or formulated, in terms of some appropriate integral equation that needs to be solved.

There are several prominent examples of this. The Karhunen-Loève theorem describes an orthonormal expansion of a random process where the set of orthonormal functions $\{\phi_i(t)\}$ is chosen such that the coefficients are uncorrelated.* These functions are specified by the solution of a homogeneous integral equation

$$\int_{T_0}^{T_f} \mathbf{K_y}(t, \tau)\boldsymbol{\phi}_i(\tau)\, d\tau = \lambda_i \boldsymbol{\phi}_i(t), \qquad T_0 \leqq t \leqq T_f, \tag{1.1}$$

where the kernel $\mathbf{K_y}(t, \tau)$ is the covariance of the expanded random process, $\boldsymbol{\phi}_i(t)$ is an eigenfunction solution, and λ_i is its associated eigenvalue, which is equal to the mean square value of the ith generalized Fourier coefficient in the expansion. In many applications this expansion simply is done conceptually in the course of an analysis. There are several problems, however, where one is interested in the actual expansion, especially the eigenvalues. In these situations obtaining specific solutions reduces to solving this homogeneous integral equation.

A problem often encountered is the detection of a known signal in the presence of a colored noise.† Typically, on one hypothesis only a noise process $\mathbf{n}(t)$ with a covariance $\mathbf{K_n}(t, \tau)$ is present, while on the other hypothesis there is a known signal $\mathbf{s}(t)$ present in addition to the noise. The optimal receiver and its performance are specified by an inhomogeneous integral equation

$$\int_{T_0}^{T_f} \mathbf{K_n}(t, \tau)\mathbf{g}(\tau)\, d\tau = \mathbf{s}(t), \qquad T_0 \leqq t \leqq T_f. \tag{1.2}$$

The receiver correlates the observed signal with the solution $\mathbf{g}(t)$ of this integral equation, and the detector performance can be related to the integrated product of $\mathbf{g}(t)$ and $\mathbf{s}(t)$. Several problems in communications essentially reduce to solving this integral equation; therefore, obtaining a solution can often be of significant practical interest.

The Wiener-Hopf equation has a fundamental importance in much of communication theory.‡ In its general form, it specifies the optimum, linear, minimum mean square error estimator of a random signal in noise. This estimator appears frequently in both detection and estimation theory problems. For example, it is the estimator in the estimator-correlator for the detection of Gaussian random signals in noise, or it is used to compute the likelihood function for estimating the parameters of a random process imbedded in noise. This equation has the form

$$\int_{T_0}^{T_f} \mathbf{h}_0(t, u)\mathbf{K}_r(u, \tau)\, du = \mathbf{K_{dr}}(t, \tau), \qquad T_0 \leqq \tau \leqq T_f, \tag{1.3}$$

* Ref. 22, pp. 96–101, Ref. 67, pp. 178–198.
† Ref. 67, pp. 287–325, or Ref. 29, pp. 95–121.
‡ Ref. 67, Chap. 6 and Ref. 68, Chaps. 3–5.

where $\mathbf{K_r}(u, \tau)$ is the covariance of an observed signal, $\mathbf{K_{dr}}(t, T)$ is the cross covariance of the desired signal and the observed signal, and $\mathbf{h}_0(t, u)$ is the impulse response of the optimum estimator at time t. Usually we are interested in the estimate rather than this impulse response; consequently, in a state variable approach we derive a differential equation structure specifying the estimate and from which the impulse response can be obtained if desired.

When T_f is fixed, this estimator corresponds to the optimum unrealizable filter, often referred to as the optimal smoother. In contrast, in the problem solved by Kalman and Bucy, T_f increases in time and t equals or exceeds T_f. Although their work has an important place in much of our discussion, we are principally concerned with the cases of fixed T_f and when t is less than an increasing T_f by a fixed amount corresponding to the optimal smoother and filter with delay, respectively.

The commonly used impulse description of systems specifies a linear integral operator, and often one is led naturally to an integral equation in the analysis of many problems. Consequently, either in the above-mentioned problems or in the context of some other, these three integral equations frequently appear in communication theory. By demonstrating how state variable concepts can be used to solve them, we can provide a useful approach to many problems that appear in communication theory.

1.2 Organization

We essentially divide the monograph into two sections. In the first, we develop from first principles the state variable solution techniques for homogeneous and inhomogeneous Fredholm integral equations. We make three essential assumptions. First, the kernel of the integral equation is a covariance function of a random process. This is the common situation in many communication theory problems. Second, a random process with this covariance function can be generated by exciting a linear system with white noise. This is analogous to generating a process with a specified spectrum by driving a system having an appropriate transfer function with white noise. Finally, we assume that the system for generating this process with the specified covariance has a known, finite dimensional state variable description of its input-output relationship. This assumption relates the description of the random processes by their covariances to their description as being generated via state variable methods.

Under these assumptions we can analyze problems involving a large class of kernels. Many kernels corresponding to covariances of the output of time-varying systems can be considered in addition to the important special case of stationary kernels, or covariances, with rational Fourier transforms corresponding to the outputs of time-invariant, or constant-parameter, systems.

To solve these integral equations, we first need to discuss how random processes propagate through linear systems described by state variables. In particular we need to develop the properties of the covariance functions of these processes. This is done in Chapter 2.

With this preliminary discussion, we study the homogeneous and inhomogeneous integral equations in Chapters 3 and 4, respectively. Both of these equations are reduced to two linear differential equations and an associated set of boundary conditions. The coefficients of these differential equations and the boundary conditions are specified directly by the matrices describing the system that generates the random process with the specified kernel.

The eigenvalues of the homogeneous equation are found to be solutions of a transcendental equation involving the transition matrix of the differential equations mentioned above. The eigenfunctions also follow directly. By using this same transcendental equation we can derive an effective method for calculating the Fredholm determinant function.

We then derive the differential equations and boundary conditions for the inhomogeneous integral equations. Since the resulting differential equations are identical with those that specify the structure of optimal smoother, we can exploit the solution techniques that have been developed in the literature for this problem. Throughout our analysis, we place our discussion in the context of the problem of detecting a known signal in the presence of colored noise.

In the second section of this monograph, we discuss two specific applications of our integral equation theory. We can observe the utility of both the actual results of the theory and the approaches used in deriving it in this context. In Chapter 5 we consider the design of optimal signals for detection in the presence of colored noise using modern optimal control theory. We focus our attention on additive signal-independent noise channels. When energy and mean square bandwidth constraints are imposed, we are able to solve the signal design problem completely. We present some specific examples, indicating both the optimal signals and performance gain over more conventional signals. While we focus on this problem, however, our approach is not limited to this class of channels. We can consider different constraints and some signal-dependent channels, including some reverberation models. To date, however, the nonlinear equations resulting from the application of the optimal control theory have not proved at all easy to solve.

In Chapter 6, we consider the estimation problems of optimum linear minimum mean square error smoothing and filtering with delay of random processes that are observed corrupted by additive noise. One of the central issues is the solution of the Wiener-Hopf equation. In the filtering problem as studied by Kalman and Bucy, the estimate is made at the end point of an observation interval using all the available past data. Since the estimate is optimum only at a single point for any particular observation interval, the filter generates what is often termed a point estimate. This observation interval increases in time as more data are received, and the filter generates a sequence of estimates, each of which is an optimum realizable, or causal, estimate of the signal process at the end point of the observation interval defined at that specific instant of time. In contrast, the smoother is a noncausal interval estimator, roughly analogous to the unrealizable filter. For a fixed observation interval, it generates an optimum estimate of the process over the entire interval. Like the Kalman-Bucy filter, the filter with delay is a point estimator with an evolving end point. However, we make an estimate at an interior point within the observation interval rather than at the end point. By allowing the delay, we can improve our estimator performance over that of the Kalman filter, and still use a realizable filter whose output evolves in time as more data are received. In both of these problems we derive the estimator structure and its associated performance. The results of our integral equation theory are the starting point for our approach.

In Chapter 7 we briefly consider some aspects of nonlinear estimation theory which can be approached using our methods. To do this we need to change our estimation criterion to one of maximum *a posteriori* probability and restrict ourselves to Gaussian processes that have been observed by means of a nonlinear modulation. Here the solution methods become rather difficult, and we are quickly led to approximate techniques.

Throughout the monograph we present many examples. We do this for two reasons. We work a number of analytic examples to illustrate the use of methods we derive. We also present a number of examples analyzed by numerical methods. In the course of the monograph we emphasize the numerical aspects of our methods. We feel this is where the major application of much of the material lies. Finding effective numerical procedures is a very relevant problem, since most problems are two complex to be analyzed analytically.

1.3 Notation

We also indicate our notational conventions. Generally, scalars are symbols in italic type, vectors are lower case symbols in boldface type, and matrices are upper case symbols in boldface type.

2 State Variable Random Processes

In this chapter we introduce the concepts and properties of state variable random processes which we need in the subsequent chapters. First, we describe the generation of random processes using state variable methods, and then we develop the properties of the second-order moments of these processes. We then indicate how this method of describing random processes compares with the conventional impulse response-covariance description, in particular for stationary processes with rational power spectra. We also introduce the description of those processes used in many of our examples. Finally, we derive a result that is common to many of the problems that we analyze.

For the reader who is familiar with random processes described by state variables, this derivation is the most important part of the chapter, for it is a key result for much of our subsequent work.

2.1 Generation of Random Processes with State Variables

In the application of the state variable methods to communication theory problems, the random processes of interest are characterized in terms of a dynamical system that is excited by a white-noise process.* Consequently, the equations describing the operation of the dynamical system and a description of the white noise excitation must be provided

* We assume that the reader is familiar with state variable methods for describing linear dynamical systems. Zadeh and Desoer,[76] DeRusso *et al.*,[24] or Athans and Falb[3] are appropriate references.

rather than the probability densities or moments of the processes. This is the point of view that we take regarding the description of our random processes.

With the exception of some of the material in Chapter 7, the processes that we consider are generated by a system whose dynamics may be described in terms of a finite-dimensional, linear, ordinary differential equation, termed the state equation:

$$d\mathbf{x}(t)/dt = \mathbf{F}(t)\mathbf{x}(t) + \mathbf{G}(t)\mathbf{u}(t) \qquad \text{(linear-state equation)}, \tag{2.1}$$

where $\mathbf{x}(t)$ is the state variable vector $(n \times 1)$, $\mathbf{u}(t)$ is the white excitation process $(m \times 1)$, and $\mathbf{F}(t)$ $(n \times n)$ and $\mathbf{G}(t)$ $(n \times m)$ are matrices that determine the system dynamics. (We work with continuous time systems. For processes that are generated by a nonlinear dynamical system, this can introduce some attendant mathematical difficulties. We are principally concerned with those generated by linear systems; consequently, we do not discuss these issues here.[61])

We assume that $\mathbf{u}(t)$ is white; i.e., it may be interpreted as the derivative of an independent increment process. Consequently, we have (assuming a zero mean)

$$E[\mathbf{u}(t)\mathbf{u}^T(\tau)] = \mathbf{Q}\delta(t - \tau). \tag{2.2}$$

In order to describe a state variable equation completely, the initial state of the system must be considered. We are concerned with representing a random process over the time interval $T_0 \leqq t \leqq T_f$. We assume that the initial state $\mathbf{x}(T_0)$ is a zero mean random vector with a covariance matrix given by

$$E[\mathbf{x}(T_0)\mathbf{x}^T(T_0)] = \mathbf{P}_0. \tag{2.3}$$

Generally, one does not observe the entire state vector at the output of a dynamical system; e.g., in many cases only the first component of the vector is observed. Consequently, we must specify the relationship between the observed random process and the state vector of the dynamic system. We assume that the observation relationship is a linear, possibly time varying, no-memory transformation such that the observed random process $\mathbf{y}(t)$ is given by

$$\mathbf{y}(t) = \mathbf{C}(t)\mathbf{x}(t) \qquad \text{(observation equation)}. \tag{2.4}$$

(If the observation is a linear transformation that involves memory, and this transformation is representable in terms of a system of state variables, i.e., there is an ordinary differential equation describing the operation, we can reduce it to the previous case by simply augmenting the state vector and then redefining the matrix $\mathbf{C}(t)$.) In Figure 2.1 we have illustrated a block diagram of the dynamic system that generates the random processes of interest.

Several comments are in order here. While this method of representing random processes is not the most general, we can discuss a large class of processes of interest. In particular, we can generate the important class of stationary processes with rational spectra quite conveniently using constant-parameter linear dynamical systems.

It is often convenient to describe the random processes in terms of the system that generates them. On occasion we do this; that is, constant-parameter systems refer to the processes that may be generated by a dynamical system with a constant-state description, e.g. stationary processes, or the Wiener process.

We have avoided introducing the assumption of Gaussian statistics for $\mathbf{x}(T_0)$ and $\mathbf{u}(t)$. We indicate whenever it is necessary to introduce this assumption; however, for many of our derivations it is unnecessary since we use a structured linear approach rather than an unstructured Gaussian approach.*

In one chapter we analyze some aspects of problems involving nonlinear modulation systems; i.e., the observation equation is nonlinear. Since the notation here is unique we defer introducing it until then.

Finally, we usually work with low-pass waveforms. In Appendix B we have introduced the concept of a complex state variable. This concept allows us to analyze bandpass waveforms of interest with very little modification to the low-pass theory that we develop in the subsequent chapters.

2.2 Covariance Functions for State Variable Processes

Many of the results of random process theory are formulated in terms of the covariance matrix for the processes of interest. In this section we show how we can relate the covariance matrix to the matrices describing the generation of the random process. We need to demonstrate two results essentially. One concerns finding $\mathbf{K}_{\mathbf{x}}(t, t)$ while the other involves the relation between $\mathbf{K}_{\mathbf{x}}(t, \tau)$ and $\mathbf{K}_{\mathbf{x}}(t, t)$.

* A structured approach constrains the form of the receiver operations allowed, while an unstructured approach assumes a more complete statistical description and realizes the structure imposed by the performance criterion. It is well known that Gaussian statistics often impose linear receiver; therefore, the two approaches often yield equivalent results. (Ref. 22, pp. 231, Ref. 67, pp. 471.)

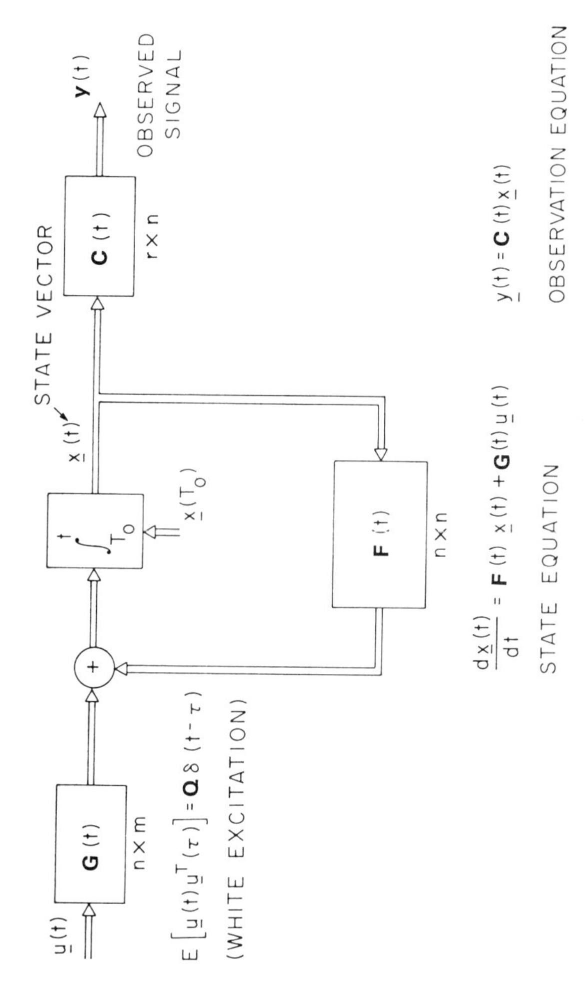

Figure 2.1 State equation model for generating $\mathbf{K_y}(t, \tau)$.

The covariance matrix of $\mathbf{y}(t)$ is defined to be (assuming a zero mean)

$$E[\mathbf{y}(t)\mathbf{y}^T(\tau)] \triangleq \mathbf{K}_\mathbf{y}(t, \tau). \tag{2.5}$$

By using Equation 2.4, $\mathbf{K}_\mathbf{y}(t, \tau)$ is easily related to the covariance matrix of the state vector $\mathbf{x}(t)$,

$$\mathbf{K}_\mathbf{y}(t, \tau) = \mathbf{C}(t)\mathbf{K}_\mathbf{x}(t, \tau)\mathbf{C}^T(\tau). \tag{2.6}$$

Since the covariance of the state vector determines the covariance of the output $\mathbf{K}_\mathbf{y}(t, \tau)$, we should focus our attention of $\mathbf{K}_\mathbf{x}(t, \tau)$.

First, we derive a differential equation for $\mathbf{K}_\mathbf{x}(t, t)$. We proceed by differentiating

$$\frac{d}{dt}\mathbf{K}_\mathbf{x}(t, t) = E\left[\frac{d\mathbf{x}(t)}{dt}\mathbf{x}^T(t) + \mathbf{x}(t)\frac{d\mathbf{x}^T(t)}{dt}\right]. \tag{2.7}$$

By substituting the state equation, we obtain

$$\frac{d}{dt}\mathbf{K}_\mathbf{x}(t, t) = \mathbf{F}(t)\mathbf{K}_\mathbf{x}(t, t) + \mathbf{K}_\mathbf{x}(t, t)\mathbf{F}^T(t)$$
$$+ \mathbf{G}(t)E[\mathbf{u}(t)\mathbf{x}^T(t)] + E[\mathbf{x}(t)\mathbf{u}^T(t)]\mathbf{G}^T(t). \tag{2.8}$$

Since the last two terms are transposes of each other, we consider only the second term. The state at time t in terms of the initial state $\mathbf{x}(T_0)$ and the input $\mathbf{u}(t')$ for $T_0 < t' < t$ is given by

$$\mathbf{x}(t) = \boldsymbol{\theta}(t, T_0)\mathbf{x}(T_0) + \int_{T_0}^{t} \boldsymbol{\theta}(t, t)\mathbf{G}(t')\mathbf{u}(t')\, dt,' \tag{2.9}$$

where $\boldsymbol{\theta}(t, \tau)$ is the transition matrix associated with $\mathbf{F}(t)$,

$$\frac{d}{dt}\boldsymbol{\theta}(t, \tau) = \mathbf{F}(t)\boldsymbol{\theta}(t, \tau), \tag{2.10a}$$

$$\boldsymbol{\theta}(\tau, \tau) = \mathbf{I}. \tag{2.10b}$$

Therefore, we have

$$E[\mathbf{x}(t)\mathbf{u}^T(t)]\mathbf{G}^T(t)$$
$$= \left\{\boldsymbol{\theta}(t, T_0)E[\mathbf{x}(T_0)\mathbf{u}^T(t)] + \int_{T_0}^{t} \boldsymbol{\theta}(t, t')\mathbf{G}(t')E[\mathbf{u}(t')\mathbf{u}^T(t)]\, dt'\right\}\mathbf{G}^T(t). \tag{2.11}$$

If we assume that the initial condition and the excitation are uncorrelated, then the first term is zero for $t > T_0$. The second term becomes, upon performing the expectation,

$$E[\mathbf{x}(t)\mathbf{u}^T(t)]\mathbf{G}^T(t) = \int_{T_0}^{t} \boldsymbol{\theta}(t, t')G(t')\mathbf{Q}\ \delta(t' - t)\ dt'\mathbf{Q}^T(t). \tag{2.12}$$

The integral is nonzero only at the end point of the integration interval. We must assume that the limiting form of the delta function is symmetrical; therefore, only one-half the "area" is included in the integration region. Integration of Equation 2.12 thus yields*

$$E[\mathbf{x}(t)\mathbf{u}^T(t)]\mathbf{G}^T(t) = \tfrac{1}{2}\mathbf{G}(t)\mathbf{Q}\mathbf{G}^T(t). \tag{2.13}$$

Substituting this term plus its transpose into Equation 2.8 gives the desired result

$$\frac{d}{dt}\mathbf{K}_{\mathbf{x}}(t, t) = \mathbf{F}(t)\mathbf{K}_{\mathbf{x}}(t, t) + \mathbf{K}_{\mathbf{x}}(t, t)\mathbf{F}^T(t) + \mathbf{G}(t)\mathbf{Q}\mathbf{G}^T(t), \qquad t > T_0. \tag{2.14}$$

The initial condition $\mathbf{K}_{\mathbf{x}}(T_0, T_0)$ must be specified, and it is given by Equation 2.3 to be $\mathbf{P}_0$.

The second result relates $\mathbf{K}_{\mathbf{x}}(t, \tau)$ to $\mathbf{K}_{\mathbf{x}}(t, t)$ and the transition matrix $\boldsymbol{\theta}(t, \tau)$. Consider the case when $t > \tau$. The state at time t is related to the state at time τ and the input $\mathbf{u}(t')$ over the interval $t > t' > \tau$ by

$$\mathbf{x}(t) = \boldsymbol{\theta}(t, \tau)\mathbf{x}(\tau) + \int_{\tau}^{t} \boldsymbol{\theta}(t, t')\mathbf{G}(t')\mathbf{u}(t')\ dt', \tag{2.15}$$

where $\boldsymbol{\theta}(t, t')$ is defined by Equation 2.10. If we postmultiply by $\mathbf{x}^T(\tau)$, and take expectations, we obtain

$$E[\mathbf{x}(t)\mathbf{x}^T(\tau)] = \boldsymbol{\theta}(t, \tau)\mathbf{K}_{\mathbf{x}}(\tau, \tau) + \int_{\tau}^{t} \boldsymbol{\theta}(t, t')\mathbf{G}(t')E[\mathbf{u}(t')\mathbf{x}^T(\tau)]\ dt'. \tag{2.16}$$

However, because of the Markov nature of the state vector, $\mathbf{u}(t')$ and $\mathbf{x}(\tau)$ are independent over the range of integration. Consequently, the second term of 2.16 is zero, and we have

$$\mathbf{K}_{\mathbf{x}}(t, \tau) = \boldsymbol{\theta}(t, \tau)\mathbf{K}_{\mathbf{x}}(\tau, \tau), \qquad t \geqq \tau, \tag{2.17}$$

which is the first part of the desired result. The derivation for the case when $\tau > t$ is identical. The net result of the two is given by

$$\mathbf{K}_{\mathbf{x}}(t, \tau) = \begin{cases} \boldsymbol{\theta}(t, \tau)\mathbf{K}_{\mathbf{x}}(\tau, \tau) & \text{for } t \geqq \tau, \\ \mathbf{K}_{\mathbf{x}}(t, t)\boldsymbol{\theta}^T(\tau, t) & \text{for } \tau \geqq t. \end{cases} \tag{2.18}$$

* The "splitting" of this impulse is a formal procedure certainly not based on distribution theory. Other approaches yield the same result for linear systems.

In the case of a deterministic input signal and deterministic initial conditions, knowledge of $\mathbf{x}(T_0)$ and $\mathbf{u}(t')$ for $T_0 \leqq t' \leqq t$ is sufficient to determine $\mathbf{x}(t)$ for all t. Similarly, with a random input and/or random initial conditions, we can also determine the covariance matrix of the state vector by first solving 2.14 and then using 2.18.

In communication theory, stationary processes are of particular interest, and it is worthwhile to comment upon where they fit in the above model. The class of stationary processes with rational power spectra can be generated by constant-parameter systems where one chooses the initial covariance matrix appropriately. (We point out that the output of a constant-parameter system need not be stationary, e.g., the Wiener process or where $\mathbf{P}_0$ is not chosen appropriately.)

If the parameters of the system generating the process $\mathbf{y}(t)$ are constant, then the transition matrix is given by the matrix exponential

$$\boldsymbol{\theta}(t, \tau) = \mathbf{e}^{\mathbf{F}(t-\tau)}. \tag{2.19}$$

We observe then that $\mathbf{K}_{\mathbf{x}}(t, t + \Delta t)$ is a function of Δt only when $\mathbf{K}_{\mathbf{x}}(t, t)$ is a constant $\mathbf{P}_\infty$. This constant matrix is the steady state solution to Equation 2.14. Consequently, we can generate a segment of a stationary process by using constant-parameter systems and setting the initial covariance $\mathbf{P}_0$ equal to $\mathbf{P}_\infty$.

By using transform techniques it is straightforward to demonstrate that the steady state solution to 2.14 is given by

$$\begin{aligned} P_\infty &= \int_0^\infty e^{\mathbf{F}t}\mathbf{GQG}e^{\mathbf{F}^Tt}\,dt \\ &= \frac{1}{2\pi j}\int_{-j\infty}^{j\infty} [\mathbf{I}s - \mathbf{F}]^{-1}\mathbf{GQG}^T[-\mathbf{I}s - \mathbf{F}^T]^{-1}\,ds. \end{aligned} \tag{2.20}$$

The state vector covariance function is

$$\mathbf{K}_{\mathbf{x}}(t + \Delta t) = \begin{cases} \mathbf{e}^{-\mathbf{F}\Delta t}\mathbf{P}_\infty, & \Delta t \leqq 0, \\ \mathbf{P}_\infty \mathbf{e}^{\mathbf{F}^T\Delta t}, & \Delta t \geqq 0. \end{cases} \tag{2.21}$$

Taking the Fourier transform yields the spectral covariance matrix

$$\mathbf{S}_{\mathbf{x}}(\omega) = [j\omega\mathbf{I} + \mathbf{F}]^{-1}\{(\mathbf{FP}_\infty) + (\mathbf{FP}_\infty)^T\}[j\omega\mathbf{I} - \mathbf{F}^T]^{-1} \tag{2.22}$$

The spectrum of the observed process, $\mathbf{S}_{\mathbf{y}}(\omega)$, is found by using Equation 2.6.

It will be useful in the text to illustrate many of our results by examples. In doing this, we will be concerned with three processes. The first is the Wiener process. A dynamical system for generating this process is

$$dx(t)/dt = u(t), \tag{2.23a}$$

$$y(t)=x(t), \tag{2.23b}$$

$$E[u(t)u(\tau)] = \sigma^2\delta(t-\tau), \tag{2.23c}$$

$$E[x^2(T_0)] = 0. \tag{2.23d}$$

The appropriate state matrices are

$$F = 0, \tag{2.24a}$$

$$G = 1, \tag{2.24b}$$

$$Q = \mu^2, \tag{2.24c}$$

$$C = 1, \tag{2.24d}$$

$$P_0 = 0. \tag{2.24e}$$

The covariance is well known to be

$$K_y(t, \tau) = \mu^2 \min(t, \tau) \qquad t, \tau \geqq 0.$$

The simplest stationary process is the first-order Butterworth process. This is the second process of interest. The state equations that describe the generation of this process are

$$dx(t)/dt = -kx(t) + u(t), \qquad t > T_0, \tag{2.25a}$$

$$y(t) = x(t), \tag{2.25b}$$

$$E[u(t)u(\tau)] = 2kP\delta(t-\tau), \tag{2.25c}$$

$$E[x^2(T_0)] = P. \tag{2.25d}$$

The matrices involved are

$$F = -k, \tag{2.26a}$$

$$G = 1, \tag{2.26b}$$

$$Q = 2kP, \tag{2.26c}$$

$$C = 1, \tag{2.26d}$$

$$P_0 = P. \tag{2.26e}$$

The covariance function for this process is

$$K_y(t, \tau) = Pe^{-k[t-\tau]}. \tag{2.27}$$

This process and the Wiener process are particularly useful in illustrating our results analytically.

Whenever the dimension of the system generating the process becomes greater than one, we generally must resort to numerical solutions. The third process of interest is generated by a two-dimensional state equation where the matrices describing the process generation are

$$\mathbf{F} = \begin{bmatrix} 0 & 1 \\ -10 & -2 \end{bmatrix}, \tag{2.28a}$$

$$\mathbf{G} = \begin{bmatrix} 0 \\ 1 \end{bmatrix}, \tag{2.28b}$$

$$Q = 160, \tag{2.28c}$$

$$\mathbf{C} = [1 \quad 0], \tag{2.28d}$$

$$\mathbf{P}_0 = \begin{bmatrix} 4 & 0 \\ 0 & 40 \end{bmatrix}. \tag{2.28e}$$

The process $\mathbf{y}(t)$ is a stationary process whose covariance function is

$$K_y(t, t + \Delta t) = \tfrac{4}{3}e^{-|\Delta t|}[3\cos(3\Delta t) + \sin(3|\Delta t|)], \tag{2.29a}$$

and whose spectrum is

$$S_y(\omega) = \frac{160}{\omega^4 - 16\omega^2 + 100}. \tag{2.29b}$$

We have illustrated these functions in Figures 2.2a and 2.2b. We have included this process principally to illustrate some of the computational aspects of our techniques; this is where several of the advantages of our methods are found. We have also chosen to state matrices such that the spectrum of $y(t)$ has a peak in it away from the origin. This introduces some interesting aspects in some of our examples. We should also note that, in all our examples, any analysis involving this process would require a prohibitive amount of time.

The principal difference between the method of characterization of systems and random processes and the more common impulse response-covariance method is that we specify the internal dynamics of the system rather than just that of the output process. We have indicated how to find the output characterization given the description of the generation method. The converse path can be difficult in general, although some recent progress in this direction has been indicated.[1] For scalar stationary processes, the problem can be solved by essentially factoring $S_y(\omega)$ and then using one of the several canonical methods for realizing the resulting transfer function.* We point out that neither the factorization (minimum phase need not be imposed) nor the resulting system be unique; consequently, several systems can be used to generate a process with the same second-order statistics. In many problems of interest,

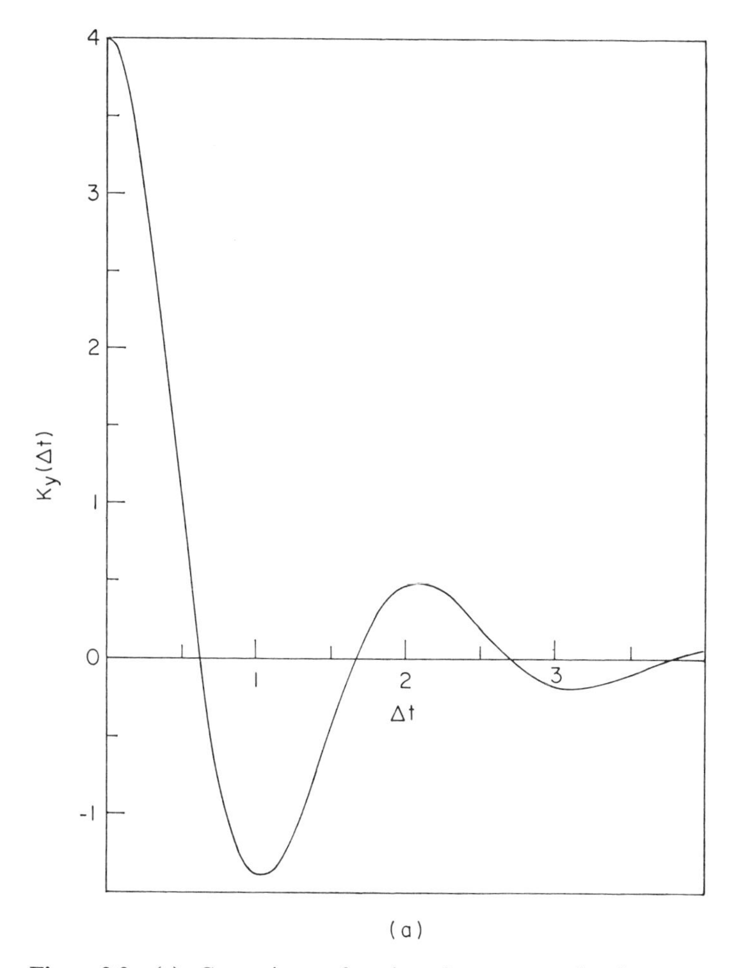

Figure 2.2 (a) Convariance function for a second-order system.

* Ref. 76, pp. 411–413.

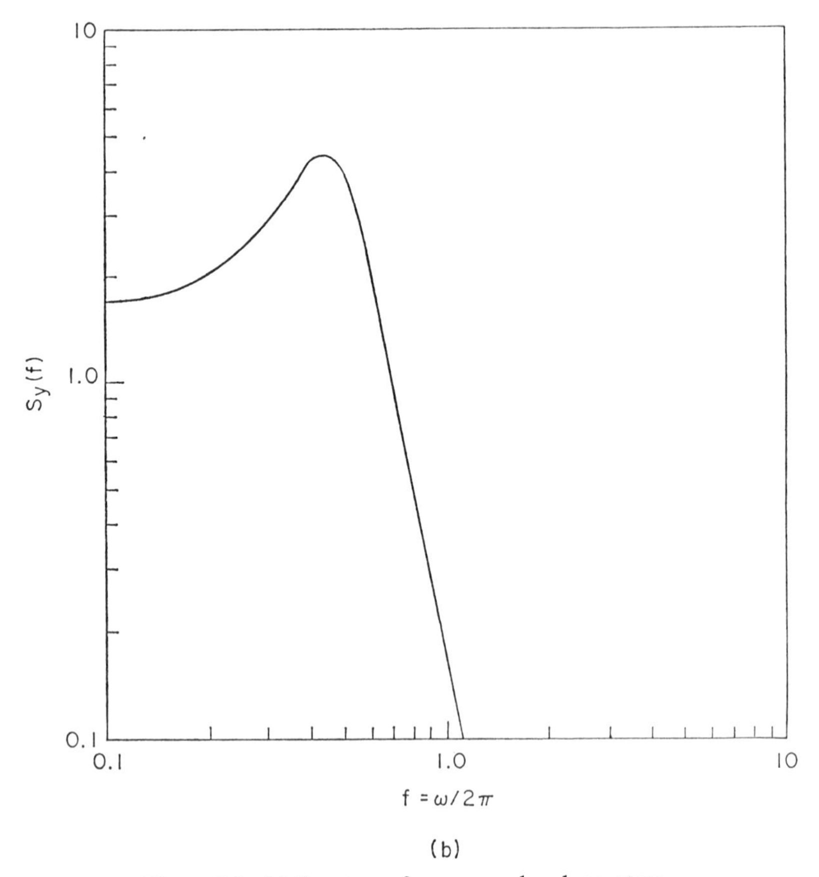

Figure 2.2 (b) Spectrum for a second-order system.

the actual covariance is only of conceptual concern since the process generation can often be inferred from the physical system involved.

2.3 The Derivation of the Differential Equations for the Covariance Operator

Many of the problems in communication theory involve the integral operation

$$\zeta(t) = \int_{T_0}^{T_f} \mathbf{K}_{\mathbf{y}}(t, \tau)\mathbf{f}(\tau)\, d\tau, \qquad T_0 \leqq t \leqq T_f, \tag{2.30}$$

In the study of Fredholm integral equations $\mathbf{f}(t)$ is either related to the eigenfunction $\boldsymbol{\phi}(t)$ in the homogeneous case, or it is the solution $\mathbf{g}(t)$ in the inhomogeneous case. In linear estimation theory, this is the integral operation specified by the Wiener-Hopf equation. In this section we derive a set of differential equations for this integral operation. Solving these differential equations is equivalent to performing the integral operation specified by Equation 2.30. In many of our derivations, we use these differential equations to convert an integral operator to an equivalent set of differential equations

Let us now proceed with our derivation. By using Equation 2.6, we may write Equation 2.30 as

$$\boldsymbol{\zeta}(t) = \mathbf{C}(t)\boldsymbol{\xi}(t), \qquad T_0 \leqq t \leqq T_f, \tag{2.31}$$

where*

$$\boldsymbol{\xi}(t) \triangleq \int_{T_0}^{T_f} \mathbf{K}_{\mathbf{x}}(t, \tau)\mathbf{C}^T(\tau)\mathbf{f}(\tau)\, d\tau, \qquad T_0 \leqq t \leqq T_f. \tag{2.32}$$

We now determine a set of differential equations in terms of the function $\boldsymbol{\xi}(t)$. Substituting Equation 2.18 in Equation 2.32, we have

$$\boldsymbol{\xi}(t) = \int_{T_0}^{t} \boldsymbol{\theta}(t, \tau)\mathbf{K}_{\mathbf{x}}(\tau, \tau)\mathbf{C}^T(\tau)\mathbf{f}(\tau)\, d\tau + \mathbf{K}_{\mathbf{x}}(t, t)\int_{t}^{T_f} \boldsymbol{\theta}^T(\tau, t)\mathbf{C}^T(\tau)\mathbf{f}(\tau)\, d\tau, \qquad T_0 \leqq t \leqq T_f. \tag{2.33}$$

If we differentiate Equation 2.33 with respect to t, we obtain

$$\frac{d\boldsymbol{\xi}(t)}{dt} = \int_{T_0}^{t} \frac{\partial \boldsymbol{\theta}(t, \tau)}{\partial t}\mathbf{K}_{\mathbf{x}}(\tau, \tau)\mathbf{C}^T(\tau)\mathbf{f}(\tau)\, d\tau + \frac{d\mathbf{K}_{\mathbf{x}}(t, t)}{dt}\int_{t}^{T_f} \boldsymbol{\theta}^T(\tau, t)\mathbf{C}^T(\tau)\mathbf{f}(\tau)\, d\tau + \mathbf{K}_{\mathbf{x}}(t, t)\int_{t}^{T_f} \frac{\partial \boldsymbol{\theta}^T(\tau, t)}{\partial t}\mathbf{C}^T(\tau)\mathbf{f}(\tau)\, d\tau, \qquad T_0 \leqq t \leqq T_f. \tag{2.34}$$

We have used Equation 2.10b and canceled two equal terms. In the first term of the right-hand side of Equation 2.34 we substitute Equation 2.10a, and in the last term we use the fact that $\boldsymbol{\theta}^T(\tau, t)$ is the transition

* Note that we could just as well derive the results for the state vector covariance operator directly. In this approach we would operate on a function $\mathbf{f}'(\tau)$ with $\mathbf{K}_{\mathbf{x}}(t, \tau)$ and then substitute $\mathbf{C}^T(\tau)\mathbf{f}(\tau)$ for the operator of interest here.

matrix for the adjoint equation of the matrix $\mathbf{F}(t)$.* That is,

$$\frac{\partial}{\partial t}\boldsymbol{\theta}^T(\tau, t) = -\mathbf{F}^T(t)\boldsymbol{\theta}^T(\tau, t). \tag{2.35}$$

When we make these two substitutions in Equation 2.34, we obtain

$$\frac{d\boldsymbol{\xi}(t)}{dt} = \mathbf{F}(t)\int_{T_0}^{t} \boldsymbol{\theta}(t, \tau)\mathbf{K}_{\mathbf{x}}(\tau, \tau)\mathbf{C}^T(\tau)\mathbf{f}(\tau)\,d\tau + \left\{\frac{d\mathbf{K}_{\mathbf{x}}(t, t)}{dt} - \mathbf{K}_{\mathbf{x}}(t, t)\mathbf{F}^T(t)\right\}\int_{t}^{T_f} \boldsymbol{\theta}^T(\tau, t)\mathbf{C}^T(\tau)\mathbf{f}(\tau)\,d\tau, \quad T_0 \leqq t \leqq T_f. \tag{2.36}$$

By applying Equation 2.14, we obtain

$$\frac{d\boldsymbol{\xi}(t)}{dt} = \mathbf{F}(t)\int_{T_0}^{t} \boldsymbol{\theta}(t, \tau)\mathbf{K}_{\mathbf{x}}(\tau, \tau)\mathbf{C}^T(\tau)\mathbf{f}(\tau)\,d\tau + [\mathbf{F}(t)\mathbf{K}_{\mathbf{x}}(t, t) + \mathbf{G}(t)\mathbf{Q}\mathbf{G}^T(t)]\int_{t}^{T_f} \boldsymbol{\theta}^T(\tau, t)\mathbf{C}^T(\tau)\mathbf{f}(\tau)\,d\tau, \quad T_0 \leqq t \leqq T_f. \tag{2.37}$$

After rearranging terms and using Equation 2.33, we finally have

$$\frac{d\boldsymbol{\xi}(t)}{dt} = \mathbf{F}(t)\boldsymbol{\xi}(t) + \mathbf{G}(t)\mathbf{Q}\mathbf{G}^T(t)\int_{t}^{T_f} \boldsymbol{\theta}^T(\tau, t)\mathbf{C}^T(\tau)\mathbf{f}(\tau)\,d\tau, \quad T_0 \leqq t \leqq T_f. \tag{2.38}$$

At this point we have derived a differential equation for $\boldsymbol{\xi}(t)$; however, we see that an integral operation still remains. Let us simply define this integral operation as a second linear functional of $\mathbf{f}(t)$,

$$\boldsymbol{\eta}(t) \triangleq \int_{t}^{T_f} \boldsymbol{\theta}^T(\tau, t)\mathbf{C}^T(\tau)\mathbf{f}(\tau)\,d\tau, \qquad T_0 \leqq t \leqq T_f. \tag{2.39}$$

Therefore, we have

$$\frac{d\boldsymbol{\xi}(t)}{dt} = \mathbf{F}(t)\boldsymbol{\xi}(t) + \mathbf{G}(t)\mathbf{Q}\mathbf{G}^T(t)\boldsymbol{\eta}(t), \qquad T_0 \leqq t \leqq T_f. \tag{2.40}$$

It is now a simple matter to derive a second differential equation that

* Ref. 76, pp. 343.

$\boldsymbol{\eta}(t)$ satisfies. Differentiating Equation 2.39 gives us

$$\frac{d\boldsymbol{\eta}(t)}{dt} + -\mathbf{C}^T(t)\mathbf{f}(t) - \mathbf{F}^T(t)\int_t^{T_f} \boldsymbol{\theta}^T(\tau, t)\mathbf{C}^T(\tau)\mathbf{f}(\tau)\,d\tau, \qquad T_0 \leqq t \leqq T_f, \tag{2.41}$$

where we have again used the adjoint relationship given by Equation 2.35. After substituting Equation 2.39, we have

$$\frac{d\boldsymbol{\eta}(t)}{dt} = -\mathbf{C}^T(t)\mathbf{f}(t) - \mathbf{F}^T(t)\boldsymbol{\eta}(t), \qquad T_0 \leqq t \leqq T_f. \tag{2.42}$$

We now want to derive two sets of boundary conditions which Equations 2.40 and 2.42 satisfy. In all the applications that we consider, the function $\mathbf{f}(\tau)$ is bounded at the end points, $t = T_0$ and $t = T_f$. Consequently, by setting $t = T_f$ in Equation 2.39, we obtain

$$\boldsymbol{\eta}(T_f) = \mathbf{0}. \tag{2.43}$$

The second boundary condition follows directly from Equation 2.33. If we set $t = T_0$, the first term is zero, while the second term may be written

$$\boldsymbol{\xi}(T_0) = \mathbf{K}_{\mathbf{x}}(T_0, T_0)\int_{T_0}^{T_f} \boldsymbol{\theta}^T(\tau, t)\mathbf{C}(\tau)\mathbf{f}(\tau)\,d\tau \tag{2.44}$$

or

$$\boldsymbol{\xi}(T_0) = \mathbf{K}_{\mathbf{x}}(T_0, T_0)\boldsymbol{\eta}(T_0) = \mathbf{P}_0\,\boldsymbol{\eta}(T_0). \tag{2.45}$$

It is easy to see that the two boundary conditions given by Equations 2.43 and 2.45 are independent.

We may now summarize the results of our derivation. We have derived two differential equations:

$$\frac{d\boldsymbol{\xi}(t)}{dt} = \mathbf{F}(t)\boldsymbol{\xi}(t) + \mathbf{G}(t)\mathbf{Q}\mathbf{G}^T(t)\boldsymbol{\eta}(t), \qquad T_0 \leqq t \leqq T_f, \tag{2.40}$$

$$\frac{d\boldsymbol{\eta}(t)}{dt} = \mathbf{C}(t)\mathbf{f}(t) - \mathbf{F}^T(t)\boldsymbol{\eta}(t), \qquad T_0 \leqq t \leqq T_f. \tag{2.42}$$

In addition, we have the boundary conditions

$$\mathbf{P}_0\,\boldsymbol{\eta}(T_0) = \boldsymbol{\xi}(T_0), \tag{2.45}$$

$$\boldsymbol{\eta}(T_f) = \mathbf{0}. \tag{2.43}$$

The relation to the original integral operation is given by

$$\boldsymbol{\zeta}(t) = \mathbf{C}(t)\boldsymbol{\xi}(t) = \int_{T_0}^{T_f} \mathbf{K}_{\mathbf{y}}(t, \tau)\mathbf{f}(\tau)\, d\tau, \qquad T_0 \leqq t \leqq T_f. \tag{2.46}$$

Notice that the only property of $\mathbf{f}(t)$ which we required was its boundedness at the end points of the interval. (This excludes considering equations of the first kind where singularity functions may appear.) Equations 2.33 and 2.45 each imply n linearly independent boundary conditions. Since the differential equations are linear, any solution that satisfies the boundary conditions is unique. Finally, the derivation of these equations can be reversed in order to obtain the functional defined by Equation 2.32; that is, we can integrate the differential equations rather than differentiate the integral equation. Consequently the solution $\boldsymbol{\xi}(t)$ to the differential equations must be identical with the result of the functional operation of Equation 2.32. This implies that the existence of a solution $\boldsymbol{\xi}(t)$ that satisfies the boundary conditions is both necessary and sufficient for the existence of the solution to the operation defined by Equation 2.32.

In this chapter we have introduced the concepts describing the generation of random processes by state variable methods. We paid particular attention to stationary processes since they are of special importance in many problems in communication theory. We also introduced three processes that we use as vehicles for illustrating our results in the subsequent chapters. We then presented a derivation reducing the covariance operator to a set of differential equations. This is a fundamental result which we often exploit. We point out, however, that it is a general derivation for representing this operator, and it need not be coupled to those specific applications that we pursue.

3 Homogeneous Fredholm Integral Equations

Homogeneous Fredholm integral equations play an important role in communication theory. As a theoretical tool, their most important use arises in the application of Karhunen-Loéve expansions for random processes. Here one expands a random process $\mathbf{y}(t)$ in an orthonormal expansion over the interval $[T_0, T_f]$*

$$\mathbf{y}(t) = \underset{N \to \infty}{\text{l.i.m.}} \sum_{i=1}^{N} y_i \boldsymbol{\phi}_i(t), \qquad T_0 \leqq t \leqq T_f, \tag{3.1}$$

where the set of generalized Fourier coefficients $\{y_i\}$ are given by

$$y_i = \int_{T_0}^{T_f} \mathbf{y}^T(t)\boldsymbol{\phi}_i(t)\,dt, \tag{3.2}$$

and the $\{\boldsymbol{\phi}_i(t)\}$ are the eigenfunction solutions to our homogeneous integral equation. While many methods have been developed for solving these equations, especially when the kernel is stationary with a rational Fourier transform, it is difficult to implement them so as to obtain actual solutions except in a limited number of cases.† References 30 and 34 contain brief surveys of some of these methods.

* We note that one can use a vector eigenfunction–scalar eigenvalue expansion. (Ref. 36; Ref. 67, pp. 220–226.) This type of expansion is necessary if the components of $\mathbf{y}(t)$ are correlated. It is useful in some control applications and in array processing methods.

† Ref. 22, pp. 371–382; Ref. 29, pp. 109, 340; Ref. 67, pp. 187–189; Ref. 45, pp. 1082–1102; Ref. 12, Secs. 7.4, 8.4, 8.5.

By applying state variable concepts to the solution of these equations, we can develop more general methods that also have the computational advantages that a state variable approach often brings. In addition, we can employ the results of our derivation to find the Fredholm determinant, which is a particularly important function in the performance analysis of many communication systems.*

3.1 The State Variable Solution to Homogeneous Fredholm Integral Equations

The homogeneous Fredholm integral equation is usually written

$$\int_{T_0}^{T_f} \mathbf{K_y}(t, \tau)\boldsymbol{\phi}(\tau)\, d\tau = \lambda\boldsymbol{\phi}(\tau), \qquad T_0 \leqq t \leqq T_f, \tag{3.3}$$

where the kernel $\mathbf{K_y}(t, \tau)$ is the covariance matrix of a vector random process $\mathbf{y}(t)$, which is generated by the methods described in the previous chapter, $\boldsymbol{\phi}(t)$ is an eigenfunction solution, and λ is the associate eigenvalue. The solution to this equation is an eigenvalue problem. There are at most a countable number of values of $\lambda > 0$ for which solutions exist to Equation 3.3, and there are no solutions for $\lambda < 0$.† If $\mathbf{K_y}(t, \tau)$ is positive definite, then the solutions have positive eigenvalues and form a complete orthonormal set. However, if $\mathbf{K_y}(t, \tau)$ is only nonnegative definite, then there exist solutions $\boldsymbol{\phi}_0(t)$ with zero eigenvalues, i.e., they are orthogonal to the kernel

$$\int_{T_0}^{T_f} \mathbf{K_y}(t, \tau)\boldsymbol{\phi}_0(\tau)\, d\tau = \mathbf{0}, \qquad T_0 \leqq t \leqq T_f. \tag{3.4}$$

We consider finding only those solutions with positive eigenvalues.

The first step in our derivation is to convert the homogeneous integral equation to a set of differential equations and boundary conditions by using the results of the previous chapters. The second step is to determine a transcendental equation for the eigenvalues by imposing the boundary conditions upon a general solution to these differential equations.

Let us put Equation 3.3 in a form where we can use our previous results. To do this we relate the kernel $\mathbf{K_y}(t, \tau)$ to the state vector

* Ref. 45, pp. 725, 731, 741, Ref. 68, Chap. 12; Refs. 20, 70.
† Ref. 4, pp. 122.

covariance matrix by using Equation 2.6. We have for Equation 3.3

$$\mathbf{C}(t)\left(\int_{T_0}^{T_f} \mathbf{K_x}(t,\tau)\mathbf{C}^T(\tau)\boldsymbol{\phi}_i(\tau)\,d\tau\right) = \lambda_i \boldsymbol{\phi}_i(t), \qquad T_0 \leqq t \leqq T_f, \quad i = 1, 2, \ldots, \tag{3.5}$$

where we have indexed the solutions with the subscript i. If in Equation 3.5 we set

$$\boldsymbol{\phi}_i(t) = \mathbf{f}(t), \tag{3.6}$$

the result is that the integral enclosed by parentheses is the function $\boldsymbol{\xi}(t)$ as defined in Equation 2.32. Consequently, let us define $\boldsymbol{\xi}_i(t)$ to be

$$\boldsymbol{\xi}_i(t) = \int_{T_0}^{T_f} \mathbf{K_x}(t,\tau)\mathbf{C}^T(\tau)\boldsymbol{\phi}_i(\tau)\,d\tau, \qquad T_0 \leqq t \leqq T_f, \tag{3.7}$$

so that Equation 3.5 becomes

$$\mathbf{C}(t)\boldsymbol{\xi}_i(t) = \lambda_i \boldsymbol{\phi}_i(t), \qquad T_0 \leqq t \leqq T_f. \tag{3.8}$$

If we assume that λ_i is positive, which is guaranteed if $\mathbf{K_y}(t,\tau)$ is positive definite, we can solve for the eigenfunction in terms of $\boldsymbol{\xi}_i(t)$. This gives us

$$\boldsymbol{\phi}_i(t) = \frac{1}{\lambda_i}\mathbf{C}(t)\boldsymbol{\xi}_i(t), \qquad T_0 \leqq t \leqq T_f, \tag{3.9}$$

If we examine Equation 3.7, we see that the integral operation which is defined is of the same form as the operation considered in Section 2.3. Consequently, we can reduce it to a set of differential equations with a two-point boundary condition. Let us identify $\boldsymbol{\phi}_i(t)$ in Equation 3.7 with $\mathbf{f}(t)$ in Equation 2.32. Then, if we substitute

$$\mathbf{f}(t) = \boldsymbol{\phi}_i(t) = \frac{1}{\lambda_i}\mathbf{C}(t)\boldsymbol{\xi}_i(t), \qquad T_0 \leqq t \leqq T_f, \tag{3.10}$$

in Equations 2.40 and 2.42, we find that these differential equations become

$$\frac{d}{dt}\boldsymbol{\xi}_i(t) = \mathbf{F}(t)\boldsymbol{\xi}_i(t) + \mathbf{G}(t)\mathbf{Q}\mathbf{G}^T(t)\boldsymbol{\eta}_i(t), \qquad T_0 \leqq t \leqq T_f, \tag{3.11}$$

$$\frac{d}{dt}\boldsymbol{\eta}_i(t) = \frac{\mathbf{C}^T(t)\mathbf{C}(t)}{\lambda_i}\boldsymbol{\xi}_i(t) - \mathbf{F}^T(t)\boldsymbol{\eta}_i(t), \qquad T_0 \leqq t \leqq T_f. \tag{3.12}$$

From Equations 2.43 and 2.45, the boundary conditions are

$$\boldsymbol{\eta}_i(T_f) = \mathbf{0}, \tag{3.13a}$$

$$\boldsymbol{\xi}_i(T_0) = \mathbf{P}_0\,\boldsymbol{\eta}_i(T_0). \tag{3.13b}$$

The desired eigenfunction is related to the solution $\boldsymbol{\xi}_i(t)$ by Equation 3.9, or

$$\boldsymbol{\phi}_i(t) = \frac{1}{\lambda_i}\mathbf{C}(t)\boldsymbol{\xi}_i(t), \qquad T_0 \leqq t \leqq T_f. \tag{3.9}$$

The net result of Equations 3.5–3.13 is that we can transform the homogeneous Fredholm integral equation into a set of differential equations and boundary conditions whose coefficients are directly related to the state equations and covariance matrices that are used to generate the random process $\mathbf{y}(t)$. The eigenvalues of the integral equation are just those values of λ_i which yield a nontrivial solution to the homogeneous equations 3.11 and 3.12 and which satisfy the homogeneous boundary conditions of 3.13.

Before we use the above results to determine the eigenvalues and eigenfunctions, let us make two observations. Notice that we have a set of $2n$ differential equations to solve. This is consistent with previous methods for treating stationary processes with rational spectra. In these methods, one has a $2n$-order differential equation to solve, where $2n$ is the degree of the denominator polynomial of the spectrum. We also remark that in general, we would not obtain the above set of state equations and boundary conditions by expressing these equations of order $2n$ in one of the canonic state representations.

Equation 3.9 implies that all of the solutions to Equation 3.3 with positive λ are contained in the range space defined by $\mathbf{C}(t)$. We should note that if $\mathbf{C}(t)$ is not "onto" for a set of t with nonzero measure, then $\mathbf{K}_{\mathbf{y}}(t, \tau)$ is not positive definite. In this situation there may be solutions with λ equal to zero which are not contained in this range space.

Our results to this point consist of a set of differential equations and boundary conditions to be satisfied. We still need to consider the problem of actually finding the eigenvalues and corresponding eigenfunctions. By demanding a nontrivial solution, we now use the above results to find a transcendental equation for the eigenvalues. Given the eigenvalues, the eigenfunctions then follow directly.

Let us define the $(2n \times 2n)$ matrix $\mathbf{W}(t : \lambda)$ as

$$\mathbf{W}(t:\lambda) = \begin{bmatrix} \mathbf{F}(t) & \mathbf{G}(t)\mathbf{Q}\mathbf{G}^T(t) \\ \dfrac{-\mathbf{C}^T(t)\mathbf{C}(t)}{\lambda} & -\mathbf{F}^T(t) \end{bmatrix}, \tag{3.14}$$

so that in vector form Equations 3.11 and 3.12 become

$$\frac{d}{dt}\begin{bmatrix} \boldsymbol{\xi}_i(t) \\ \boldsymbol{\eta}_i(t) \end{bmatrix} = \mathbf{W}(t:\lambda_i)\begin{bmatrix} \boldsymbol{\xi}_i(t) \\ \boldsymbol{\eta}_i(t) \end{bmatrix}, \qquad T_0 \leqq t \leqq T_f. \tag{3.15}$$

Furthermore, let us define the transition matrix associated with $\mathbf{W}(t:\lambda)$ by

$$\frac{\partial}{\partial t}\mathbf{\Psi}(t, T_0:\lambda) = \mathbf{W}(t:\lambda)\mathbf{\Psi}(t, T_0:\lambda), \tag{3.16}$$

$$\mathbf{\Psi}(T_0, T_0:\lambda) = \mathbf{I}. \tag{3.17}$$

We emphasize the λ dependence of $\mathbf{W}(t:\lambda)$ and $\mathbf{\Psi}(t, T_0:\lambda)$ by including λ as an argument.

In terms of this transition matrix, the most general solution to Equation 3.14 is

$$\begin{bmatrix}\boldsymbol{\xi}_i(t)\\ \boldsymbol{\eta}_i(t)\end{bmatrix} = \mathbf{\Psi}(t, T_0:\lambda_i)\begin{bmatrix}\boldsymbol{\xi}_i(T_0)\\ \boldsymbol{\eta}_i(T_0)\end{bmatrix}, \qquad T_0 \leqq t \leqq T_f. \tag{3.18}$$

After employing the boundary condition specified by Equation 3.13, we have

$$\begin{bmatrix}\boldsymbol{\xi}_i(t)\\ \boldsymbol{\eta}_i(t)\end{bmatrix} = \mathbf{\Psi}(t, T_0:\lambda_i)\begin{bmatrix}\mathbf{P}_0\\ \mathbf{I}\end{bmatrix}\boldsymbol{\eta}_i(T_0), \qquad T_0 \leqq t \leqq T_f. \tag{3.19}$$

Let us now partition $\mathbf{\Psi}(t, T_0:\lambda)$ into four n by n matrices such that

$$\mathbf{\Psi}(t, T_0:\lambda) = \begin{bmatrix}\mathbf{\Psi}_{\xi\xi}(t, T_0:\lambda) & \mathbf{\Psi}_{\xi\eta}(t, T_0:\lambda)\\ \mathbf{\Psi}_{\eta\xi}(t, T_0:\lambda) & \mathbf{\Psi}_{\eta\eta}(t, T_0:\lambda)\end{bmatrix}. \tag{3.20}$$

Rewriting Equation 3.19 in terms of these partitions we have

$$\begin{bmatrix}\boldsymbol{\xi}_i(t)\\ \boldsymbol{\eta}_i(t)\end{bmatrix} = \begin{bmatrix}\mathbf{\Phi}_{\xi}(t, T_0:\lambda_i)\\ \mathbf{\Phi}_{\eta}(t, T_0:\lambda_i)\end{bmatrix}\boldsymbol{\eta}_i(T_0), \qquad T_0 \leqq t \leqq T_f, \tag{3.21}$$

where we define

$$\mathbf{\Phi}_{\xi}(t, T_0:\lambda) = \mathbf{\Psi}_{\xi\xi}(t, T_0:\lambda_i)\mathbf{P}_0 + \mathbf{\Psi}_{\xi\eta}(t, T_0:\lambda_i), \tag{3.22a}$$

$$\mathbf{\Phi}_{\eta}(t, T_0:\lambda) = \mathbf{\Psi}_{\eta\xi}(t, T_0:\lambda_i)\mathbf{P}_0 + \mathbf{\Psi}_{\eta\eta}(t, T_0:\lambda_i). \tag{3.22b}$$

Note that $\mathbf{\Phi}_{\xi}(t, T_0:\lambda)$ and $\mathbf{\Phi}_{\eta}(t, T_0:\lambda)$ satisfy 3.15 with the initial condition $\mathbf{P}_0$ and $\mathbf{I}$, respectively. Taking the lower partition, the boundary condition given by Equation 3.13b requires

$$\mathbf{0} = \boldsymbol{\eta}_i(T_f) = \mathbf{\Phi}_{\eta}(T_f, T_0:\lambda_i)\boldsymbol{\eta}_i(T_0). \tag{3.23}$$

This implies one of two consequences. Either we have $\boldsymbol{\eta}_i(T_0)$ identically zero, which leads to the trivial zero solution, or

$$\det[\mathbf{\Phi}_{\eta}(T_f, T_0:\lambda_i)] = 0. \tag{3.24}$$

If the latter is true, Equation 3.15 has a nontrivial solution that satisfies the requisite boundary conditions. Because of the functional equivalence of these differential equations and the original integral equation, this nontrivial solution to Equation 3.15 implies that λ_i is an eigenvalue. That is, the eigenvalues of Equation 3.3 are simply those of values of λ_i that satisfy the transcendental equation specified by Equation 3.24.

Let us briefly review our results by suggesting an algorithm for obtaining the eigenvalues and eigenfunctions for the homogeneous integral equation. In order to find the eigenvalues we need to find the solutions to Equation 3.24. To do this we compute the transition matrix $\mathbf{\Psi}(t, T_0 : \lambda)$ as given by 3.16 and evaluate it at $t = T_f$. We then partition this matrix as indicated by Equation 3.20 and form the matrix $\mathbf{\Phi}_{\boldsymbol{\eta}}(T_f, T_0 : \lambda)$. Let us define the function

$$A(\lambda) = \det[\mathbf{\Phi}_{\boldsymbol{\eta}}(T_f, T_0 : \lambda)]. \tag{3.25}$$

To find the eigenvalues we look for the solutions to the equation $A(\lambda) = 0$ as specified by Equation 3.24.

In the cases where we have an analytic expression for $A(\lambda)$ we can solve for its roots directly. In those cases where we evaluate the function $A(\lambda)$ numerically, we must plot this function versus λ and locate its zero crossings. In actually plotting this function it is useful to have an upper bound to the largest eigenvalue, so that one knows where to begin. We point out that a convenient bound is given by the expected value of the energy in the process, as

$$\lambda_{\max} \leqq \sum_{i=1}^{\infty} \lambda_i = \int_{T_0}^{T_f} \operatorname{Tr}[\mathbf{K}_{\mathbf{y}}(t, t)]\, dt = E. \tag{3.26}$$

If the kernel is stationary and a scalar, we can also bound $\lambda_{\max}$ by the maximum value of the corresponding spectrum $S_y(f)$.* Similar results hold for matrix kernels and the trace of $\mathbf{S}_{\mathbf{y}}(f)$.

In order to determine the associated eigenfunction $\boldsymbol{\phi}_i(t)$, we note that Equation 3.24 implies that the characteristic polynomial of $\mathbf{\Phi}_{\boldsymbol{\eta}}(T_f, T_0 : \lambda_i)$ has a root equal to zero and $\boldsymbol{\eta}_i(T_0)$ is the characteristic vector associated with this root. (We have used the adjective "characteristic" in order to avoid confusing the eigenvalue properties of the *matrix* $\mathbf{\Phi}_{\boldsymbol{\eta}}(T_f, T_0 : \lambda_i)$ with those of the integral equation, Equation 3.3.) Therefore, to determine $\boldsymbol{\eta}_i(T_0)$ to a multiplicative factor we solve the linear homogeneous equation

$$\mathbf{\Phi}_{\boldsymbol{\eta}}(T_f, T_0 : \lambda_i)\boldsymbol{\eta}_i(T_0) = \mathbf{0}. \tag{3.27}$$

* Ref. 67, pp. 208.

Given $\boldsymbol{\eta}_i(T_0)$ we can find the eigenfunctions by using the upper partition of Equation 3.8 and Equation 3.22a. This gives us

$$\boldsymbol{\phi}_i(t) = \frac{\mathbf{C}(t)}{\lambda_i} \boldsymbol{\Phi}_{\xi}(t, T_0 : \lambda_i)\boldsymbol{\eta}_i(T_0), \qquad T_0 \leqq t \leqq T_f, \tag{3.28}$$

which is the desired result.

Before proceeding we should comment about multiple-order roots of Equation 3.24. In general, the function $A(\lambda)$ vanishes with nonzero slope, that is, near an eigenvalue λ_i,

$$A(\lambda) = c_1(\lambda - \lambda_i) + c_2(\lambda - \lambda_i)^2 + \cdots, \tag{3.29}$$

where c_1 is nonzero. In the case of multiple-order eigenvalues, the function $A(\lambda)$ vanishes tangentially; that is, near an eigenvalue λ_i of order l

$$A(\lambda) = c_l(\lambda - \lambda_i)^l + c_{l+1}(\lambda - \lambda_i)^{l+1} + \cdots. \tag{3.30}$$

This implies that there will be l linearly independent vectors $\boldsymbol{\eta}_i^l(T_0)$ satisfying

$$\boldsymbol{\Phi}_{\eta}(T_f, T_0 : \lambda_i)\boldsymbol{\eta}_i^l(T_0) = \mathbf{0}; \tag{3.31}$$

i.e., $\boldsymbol{\Phi}_{\eta}(T_f, T_0 : \lambda_i)$ has rank $n - l$. Note that the dimension of the state vector sets an upper bound on the multiplicity of any eigenvalue since $\eta - l \geqq 0$, or $\eta \geqq l$.

Before proceeding, let us point out some advantages that this technique offers.

1. Once the state matrices are chosen to generate $\mathbf{y}(t)$, the differential equations that must be solved follow directly. The reduction of the integral equation to the differential equations and boundary conditions is virtually automatic.

2. We can solve for each eigenvalue and eigenfunction independently of the others, which is significant when actually obtaining accurate solutions.

3. One does not have to substitute any function back into the original integral equation in order to match coefficients and to determine the transcendental equation that determines the eigenvalues.

4. We can obtain solutions in the vector case. For those techniques that rely upon spectral factorization methods the vector case could cause some difficulty. (In some respects we have defined this problem away by our method of characterizing the random processes of interest. It should be pointed out that, depending upon the problem, this method of characterization may be just as fundamental as the covariance method.)

5. We can study a certain class of time-varying kernels by using numerical methods, e.g., those corresponding to models of spread channels.

6. Finally, the most important advantage is that the technique is well suited to numerical methods. This allows one to determine numerical solutions easily for problems in which an analytic calculation is either difficult or not feasible.

Let us expand briefly upon this last advantage. Generally, for systems of dimension two or more, the algebraic tedium of any method makes a computational algorithm highly desirable. Upon this issue, we direct some comments in two general directions.

The first concerns the implementation of one of the several algorithms suggested for stationary kernels in references cited at the beginning of the chapter. Determining the trial solutions as a function of λ which are usually required is not especially difficult, although it may require a numerical solution for the roots of the polynomial involved. The substitution of these solutions back into the original integral equation to find the transcendental equation for the eigenvalues is difficult and awkward even with numerical techniques.

In the proposed method there is really only one fundamental calculation required, the determination of the transition matrix $\mathbf{\Psi}(t, T_0 : \lambda)$. For stationary kernels this is especially easy because the system parameters for generating the kernel are constants; consequently the transition matrix can be evaluated by a matrix exponential

$$\mathbf{\Psi}(t, T_0 ; \lambda) = \mathbf{e}^{\mathbf{W}(\lambda)(t-T_0)} . \tag{3.32}$$

Although one can do this analytically using Laplace transform techniques,*

$$\mathbf{e}^{\mathbf{W}(\lambda)t} = \mathscr{L}^{-1}\{[\mathbf{I}s - \mathbf{W}(\lambda)]^{-1}\}, \tag{3.33}$$

where $\mathscr{L}^{-1}$ is the inverse Laplace operator, it is more important to consider its numerical evaluation. Several methods have been proposed in the recent literature. Some evolve around the truncation of the series for the matrix exponential,

$$\mathbf{e}^{\mathbf{W}(\lambda)t} = \sum_{j=0}^{\infty} \frac{[\mathbf{W}(\lambda)]^j t^j}{j!} . \tag{3.34}$$

Others involve the use of the so-called Faddeeva algorithm (see Appendix A).[43] This algorithm is particularly noteworthy because of its

* (In the inversion the contour must be taken to the right of all pole locations of $[\mathbf{I}s - W(\lambda)]^{-1}$.)

rapid execution time. In any case we can focus our attention on the computation of this function since our boundary conditions are explicit and no resubstitution is required.

The second direction concerns methods that essentially convert the integral equation into an algebraic equation. The major difficulty here is the large dimension of the matrices involved. For example, in a recent paper matrices of dimension of 100×100 were involved for accurate results.[72] Other methods are available: for example, convergence on the largest eigenvalue and then reducing the kernel, i.e., Kellog's method.* Here, obtaining accurate solutions again requires a large number of calculations because of the dimensionality involved. Obviously, we have not considered all the available methods, nor have we done an extensive comparison. We have found the proposed method to be very accurate and to require small amounts of computer time even for systems of fifth order.

There are two disadvantages:

1. If the kernel is given, the determination of state matrices for describing its generation can be a difficult problem in general. Quite often, the state variable description is the more fundamental characterization. (Some recent work directed to this problem has appeared.[1])
2. The class of kernels that we can analyze is limited. Most processes of interest in communication theory, however, do fit within our model.

3.2 The Fredholm Determinant Function

In the application of the Fredholm theory to communication theory problems, the Fredholm determinant function is often used, e.g. in the design of receivers and the calculation of their performance of detecting Gaussian signals in Gaussian noise.† This function is defined as

$$D_{\mathscr{F}}(z) = \prod_{i=1} (1 + z\lambda_i), \tag{3.35}$$

where the λ_i are the eigenvalues of the homogeneous integral equation. In this section we indicate how the theory developed in Section 3.1 can be used to find a closed-form expression for this function.

Let us consider the function $A(-1/z)$. We observe that, because the eigenvalue test of the previous section is both necessary and sufficient, this function has the same set of zeroes with the same multiplicity as

* Ref. 31, pp. 156.
† Ref. 68, Chap. 3.

the Fredholm determinant. Consequently, it is logical to assert that it is proportional to the Fredholm determinant.*

To prove this assertion, we need to use a result from function theory. First $A(-1/z)$ is analytic in the finite plane. Next we observe that the dependence of $\mathbf{W}(t:-1/z)$ upon z is first order; therefore, each term in the transition matrix $\mathbf{\Psi}(t, T_0:-1/z)$ can be bounded by $\exp[|z|^{1+\varepsilon}]$ for all $\varepsilon > 0$. For constant-parameter systems, this is easy to see because of the matrix exponential nature of the transition matrix. For time-varying systems, we can use a series of piecewise constant approximations and the semigroup property of transition matrices to prove this. Since the determinant operation represents an algebraic operation, we can bound $A(-1/z)$ in the same manner. Consequently, we can demonstrate that the order, ρ, of $A(-1/z)$ is less than or equal to one.† Given this, one can prove that $A(-1/z)$ has the form‡

$$A(-1/z) = e^{az+b}D_{\mathscr{F}}(z). \tag{3.36}$$

To determine a and b we consider them to be functions of the end point T_f. We have

$$\frac{\partial}{\partial T_f}\ln A(-1/z:T_f)\bigg|_{z=0} = \dot{b}(T_f), \tag{3.37}$$

and

$$\begin{aligned}\frac{\partial^2}{\partial z\,\partial T_f}\ln A(-1/z:T_f)\bigg|_{z=0} &= \frac{d}{dT_f}\left[a(T_f) + \sum_{i=1}^{\infty}\lambda_i(T_f)\right]\\ &= \dot{a}(T_f) + \dot{E}(T_f)\\ &= \frac{da(T_f)}{dT_f} + \mathrm{Tr}[\mathbf{K}_{\mathbf{y}}(T_f, T_f)].\end{aligned} \tag{3.38}$$

We have

$$\begin{aligned}&\frac{\partial}{\partial T_f}\ln A(-1/z:T_f)\\ &\quad= \frac{\partial}{\partial T_f}\ln\det[\mathbf{\Phi}_{\boldsymbol{\eta}}(T_f, T_0:-1/z)]\\ &\quad= \mathrm{Tr}\left\{\mathbf{\Phi}_{\boldsymbol{\eta}}^{-1T}(T_f, T_0:-1/z)\frac{\partial}{\partial T_f}\mathbf{\Phi}_{\boldsymbol{\eta}}^T(T_f, T_0:-1/z)\right\}\\ &\quad= z\,\mathrm{Tr}\left\{\mathbf{C}^T(T_f)\mathbf{C}(T_f)\mathbf{\Phi}_{\boldsymbol{\xi}}(T_f, T_0:-1/z)\mathbf{\Phi}_{\boldsymbol{\eta}}^{-1}(T_f, T_0:-1/z)\right\}\\ &\qquad\qquad - \mathrm{Tr}\,\mathbf{F}^T(T_f),\end{aligned} \tag{3.39}$$

* This was asserted in Ref. 7. Later Collins proved it in Ref. 19 using a result connected with the realizable filter theory.

† Ref. 64, pp. 248.

‡ Ref. 64, pp. 250.

where we have used a matrix identity in Ref. 4 and then Equations 3.22 and 3.15. Therefore, we next obtain from 3.37 and 3.39

$$\frac{d}{dT_f} b(T_f) = \frac{\partial}{\partial T_f} \ln A(-1/z : T_f)\bigg|_{z=0} = -\mathrm{Tr}[\mathbf{F}^T(T_f)]; \tag{3.40}$$

and from Equation 3.38

$$\frac{da(T_f)}{dT_f} + \mathrm{Tr}[\mathbf{K}_\mathbf{y}(T_f, T_f)] = \frac{\partial^2}{\partial z\, \partial T_f} \ln A(-1/z : T_f)\bigg|_{z=0}$$

$$= \mathrm{Tr}\left\{\mathbf{C}(T_f)\left([\mathbf{\Phi}_\xi(T_f, T_0 : -1/z)\mathbf{\Phi}_\eta^{-1}(T_f, T_0 : -1/z)]\bigg|_{z=0}\right)\mathbf{C}^T(T_f)\right\}. \tag{3.41}$$

We have that

$$\frac{d}{dT_f}(\mathbf{\Phi}_\xi(T_f, T_0 : -1/z)\mathbf{\Phi}_\eta^{-1}(T_f, T_0 : -1/z))\bigg|_{z=0}$$

$$= (\mathbf{F}(T_f)\mathbf{\Phi}_\xi(T_f, T_0 : -1/z) + \mathbf{G}(T_f)\mathbf{Q}\mathbf{G}^T(T_f)\mathbf{\Phi}_\eta(T_f, T_0 : -1/z))$$
$$\mathbf{\Phi}_\eta^{-1}(T_f, T_0 : -1/z))|_{z=0}$$
$$- (\mathbf{\Phi}_\xi(T_f, T_0 : -1/z)\mathbf{\Phi}_\eta^{-1}(T_f, T_0 : -1/z)$$
$$(z\mathbf{C}^T(t)\mathbf{C}(t)\mathbf{\Phi}_\xi T_f, T_0 : -1/z) - \mathbf{F}^T(T_f)\mathbf{\Phi}_\eta(T_f, T_0 : -1/z))$$
$$\mathbf{\Phi}_\eta^{-1}(T_f, T_0 : 1/z)|_{z=0}$$
$$= \mathbf{F}(T_f)\mathbf{\Phi}_\xi(T_f, T_0 : -1/z)\mathbf{\Phi}_\eta^{-1}(T_f, T_0 : -1/z))\bigg|_{z=0}$$
$$+ (\mathbf{\Phi}_\xi(T_f \cdot T_{0\,\lambda} -1/z)\mathbf{\Phi}_\eta^{-1}(T_f, T_0 : -1/z))\bigg|_{z=0} \mathbf{F}^T(T_f)$$
$$+ \mathbf{G}(T_f)\mathbf{Q}\mathbf{G}^T(T_f). \tag{3.42}$$

This is the same differential equation as the unobserved variance equation, Equation 2.14. Since the equation is linear and the initial conditions are identical, we have that*

$$\mathbf{\Phi}_\xi(T_f, T_0 : -1/z)\mathbf{\Phi}_\eta^{-1}(T_f, T_0 : -1/z)\bigg|_{z=0} = \mathbf{K}_\mathbf{x}(T_f, T_f). \tag{3.43}$$

Consequently, by using 3.41, 3.43, and 2.6 we find that

$$\frac{d}{dT_f} a(T_f) = 0. \tag{3.44}$$

* For positive z we obtain the Ricatti equation for the realizable filter. This result is developed and used in an analysis of the inhomogeneous equation in Chap. 4.

When T_f equals T_0, $A(\lambda)$ obviously is zero; therefore, $a(T_f) = 0$. We finally have, by integrating 3.40 and substituting in 3.36,

$$A(-1/z) = \exp\left(-\int_{T_0}^{T_f} \text{Tr}[\mathbf{F}^T(t)]\, dt\right) D_{\mathscr{F}}(z) \tag{3.45}$$

or

$$D_{\mathscr{F}}(z) = \exp\left(\int_{T_0}^{T_f} \text{Tr}[\mathbf{F}^T(t)]\, dt\right)\left[A\left(-\frac{1}{z}\right)\right].^{*} \tag{3.46}$$

For many of the applications of the Fredholm determinant function to communication theory problems. $D_{\mathscr{F}}(z)$ is not the most convenient function to use. Let us introduce a related function

$$d(\lambda) = D_{\mathscr{F}}\left(\frac{1}{\lambda}\right) = \prod_{i=1}^{\infty}\left(1 + \frac{\lambda_i}{\lambda}\right) = \exp\left(\int_{T_0}^{T_f} \text{Tr}[\mathbf{F}^T(t)]\, dt\right) A(-\lambda). \tag{3.47}$$

This function approaches unity as $\lambda \to \pm\infty$. For positive values of λ, it is a monotonically decreasing function of λ, while for negative values of λ, it has the same behavior as the function that determines the eigenvalues λ_i. The essential conclusion is that we can evaluate the Fredholm determinant function $D_{\mathscr{F}}(z)$ by using the same function that we employed for determining the eigenvalues of the integral equation except that we use the argument $-1/z$ for λ.

Before proceeding, we also mention an asymptotic expression for $d(\lambda)$ when $\mathbf{y}(t)$ is a stationary process and the time interval, or $2WT$ product, is large. Under these assumptions it is easy to show that,†[4]

$$\ln d(\lambda) \cong T \int_{-\infty}^{\infty} \ln \det\left(\mathbf{I} + \frac{\mathbf{S}_y(\omega)}{\lambda}\right)\frac{d\omega}{2\pi}. \tag{3.48}$$

This formula allows us to determine the asymptotic behavior of $d(\lambda)$ in our calculations.

* Our result is analogous to that derived by Siegert in his studies of the characteristic function of stationary, Markov, Gaussian noise.[56] He also has related the Fredholm determinant to the solution of an inhomogeneous Fredholm integral equation.[57] With the introduction of state variable methods these techniques may be equivalent to ours in implementation difficulty from a computational viewpoint.

† The key to this analysis is to note that in distribution the eigenvalues approach the spectrum evaluated at the harmonics of $1/T$. Therefore we obtain

$$\ln d(\lambda) = \sum_{i=1}^{\infty} \ln \det\left(1 + \frac{\lambda_i}{\lambda}\right) \cong T \sum_i \ln \det\left(\mathbf{I} + \frac{\mathbf{S}(f_i)}{\lambda}\right)\frac{1}{T} \cong T\int_{-\infty}^{\infty} \ln \det\left(\mathbf{I} + \frac{\mathbf{S}(f)}{\lambda}\right) df.$$

See Ref. 67, pp. 207.

3.3 Examples

In this section we illustrate the methods developed in the two previous sections. We consider several examples. First, we do three examples analytically, principally to illustrate the use of the formulas. These processes are generated by first-order systems, which are the only type that can be analyzed in a reasonable amount of time. We then analyze processes generated by higher-order systems by numerical methods. It is this type of problem where the technique is most useful, for it allows one to obtain numerical solutions very quickly with a digital computer.

Example 3.1 The Wiener Process

The simplest process that can be generated using state variable methods is the Wiener process. It can be analyzed very simply with our methods.* The covariance matrix of a Wiener process which starts at $t = 0$ is

$$K_y(t, \tau) = \mu^2 \min(t, \tau), \qquad 0 \leqq t, \tau. \tag{3.49}$$

A state variable representation of a system that generates $y(t)$ is given by Equations 2.23 and 2.24.

Let us find the solution to Equation 3.3 when we choose $T_f = T$ and $T_0 = 0$. First, we need the matrix $\mathbf{W}(\lambda)$ (the system is constant parameter) specified by Equation 3.14. After performing the required substitution of Equation 2.24 in Equation 3.14 we have

$$\mathbf{W}(\lambda) = \begin{bmatrix} 0 & 1 \\ -\dfrac{\mu^2}{\lambda} & 0 \end{bmatrix}. \tag{3.50}$$

To find the eigenvalues and eigenfunctions we need to find the transition matrix of $\mathbf{W}(\lambda)$. If we apply Equation 3.33 we find that the transition matrix $\mathbf{\Psi}(t, 0 : \lambda)$ is

$$\mathbf{\Psi}(t, 0 : \lambda) = \begin{pmatrix} \cos\left(\dfrac{\mu}{\sqrt{\lambda}} t\right) & \dfrac{\sqrt{\lambda}}{\mu} \sin\left(\dfrac{\mu}{\sqrt{\lambda}} t\right) \\ -\dfrac{\mu}{\sqrt{\lambda}} \sin\left(\dfrac{\mu}{\sqrt{\lambda}} t\right) & \cos\left(\dfrac{\mu}{\sqrt{\lambda}} t\right) \end{pmatrix}. \tag{3.51}$$

* It is both interesting and useful to contrast our analysis of this example to those developed elsewhere. The advantages discussed earlier quickly become apparent.

We now simply apply the results as summarized at the end of the previous section.

First we substitute Equations 2.24 and 3.51 evaluated at $t = T$ into Equations 3.22 and 3.25 to find $A(\lambda)$:

$$A(\lambda) = \cos\left(\frac{\mu}{\sqrt{\lambda}} T\right). \tag{3.52}$$

The eigenvalues are given by the roots of $A(\lambda_i) = 0$, as specified by Equation 3.24. The distinct roots are given by

$$\lambda_i = \left\{\frac{2\mu T}{(2i+1)\pi}\right\}^2, \tag{3.53}$$

$$i = 0, 1, 2, \ldots.$$

The eigenfunctions follow by substituting Equation 3.37 in Equation 3.26. After determining the appropriate normalization factor, we have

$$\phi_i(t) = \sqrt{\frac{2}{T}} \sin\left(\left(\frac{2i+1}{2}\right)\frac{\pi}{T} t\right), \qquad 0 \leqq t \leqq T. \tag{3.54}$$

To find the Fredholm determinant, we first note that $F(t) = 0$; therefore, the proportionality factor is unity. Using Equation 3.53 in Equation 3.46 we have

$$D_{\mathscr{F}}(z) = \cosh(\mu\sqrt{z}\, T) \tag{3.55}$$

or

$$d(\lambda) = \cosh\left(\frac{\mu}{\sqrt{\lambda}} T\right). \tag{3.56}$$

Example 3.2a First-Order Stationary Process: The First-Order Butterworth

Let us now consider the kernel of Equation 3.3 to be

$$K_y(t, \tau) = Pe^{-k|t-\tau|}. \tag{3.57}$$

This is the covariance of the output of a first-order system with a pole at $-k$ and P_0 chosen such that the process is stationary. The state equations that generate this process are given by Equations 2.25 and 2.26. Since the kernel is stationary, only the difference between the upper and lower limits of the integral are important. Consequently, we again set $T_f = T$ and $T = 0$.

Proceeding as before, the matrix $\mathbf{W}(\lambda)$ follows by substituting Equations 2.26 and 3.14.

$$\mathbf{W}(\lambda) = \begin{bmatrix} -k & 2kP \\ -\dfrac{1}{\lambda} & k \end{bmatrix}. \tag{3.58}$$

The transition matrix for $\mathbf{W}(\lambda)$ is

$$\mathbf{\Psi}(t, 0 : \lambda) = \begin{pmatrix} \cos(kbt) - \dfrac{\sin(kbt)}{b} & \dfrac{2P}{b}\sin(kbt) \\ -\dfrac{1}{\lambda k}\dfrac{\sin(kbt)}{b} & \cos(kbt) + \dfrac{\sin(kbt)}{b} \end{pmatrix}, \tag{3.59}$$

where

$$b \triangleq \sqrt{\frac{2P}{k\lambda} - 1}. \tag{3.60}$$

By substituting Equation 3.43 in 3.25,

$$A(\lambda) = \frac{1}{b}\left[1 - \frac{P}{\lambda k}\right]\sin(kbT) + \cos(kbT). \tag{3.61}$$

The eigenvalues are given by the roots of this equation. In order to compute the roots by hand, these roots can be found by solving the equation

$$\tan(kb_i T) = \frac{2b_i}{b_i^2 - 1}. \tag{3.62}$$

Solving Equation 3.62 for λ gives us the expression for the eigenvalues λ_i in terms of the b_i,

$$\lambda_i = \frac{2P}{k}[1 + b_i^2]^{-1}. \tag{3.63}$$

Applying Equation 3.28 gives us the eigenfunctions. They are of the form

$$\phi_i(t) = \gamma_i\left[\cos(kb_i t) + \frac{1}{b_i}\sin(kb_i t)\right], \qquad 0 \leqq t \leqq T, \tag{3.64}$$

where γ_i is a normalizing factor.

Since $F(t) = -k$, the proportionality factor for the Fredholm determinant is given by $\exp(-kT)$. To determine $A(-\lambda)$, we use Equation 3.59. Let us define

$$\beta = \left[1 + \frac{2P}{k\lambda}\right]^{1/2}. \tag{3.65}$$

After some routine algebra, we obtain

$$\begin{aligned} d(\lambda) &= e^{-kT}\left[\left(\frac{\beta^2+1}{2\beta}\right)\sinh(k\beta T) + \cosh(k\beta T)\right] \\ &= \frac{e^{-kT}}{2\beta}\left[(\beta+1)^2 e^{k\beta T} - (\beta-1)^2 e^{-k\beta T}\right]. \end{aligned}$$

Example 3.2b A First-Order Nonstationary Process*

The output process $\mathbf{y}(t)$ of a constant-parameter system is not necessarily, for example, the Wiener process. A second illustration of this can be generated from the previous example. Instead of setting the initial state error covariance, P_0, equal to the mean-square power of the stationary process, let us assume that we know the initial status at $t = 0$ exactly; i.e.,

$$P_0 = 0.$$

For this situation, the kernel $K_y(t, \tau)$ becomes

$$K_y(t, \tau) = \begin{cases} Pe^{-kt}(e^{k\tau} - 1), & t > \tau, \\ Pe^{-k\tau}(e^{kt} - 1), & \tau \geqq t. \end{cases} \tag{3.66}$$

Proceeding exactly as before for the stationary case, we find that the equation for the b_i now becomes

$$\tan(kb_i T) = -b_i,$$

and the eigenfunctions have the form

$$\phi_i(t) = \gamma_i \sin(kb_i t), \qquad 0 \leqq t \leqq T, \tag{3.67}$$

where γ_i is again a normalization factor. The function $d(\lambda)$ also follows from 3.59:

$$d(\lambda) = e^{kT}\left(\cosh(k\beta T) + \frac{\sinh(k\beta T)}{\beta}\right) \tag{3.68}$$

* Kailath has analyzed several examples of integral equations with nonstationary and/or nonrational kernels.[34]

Example 3.3 A Second-Order Stationary Process

In this example we want to consider the analysis when the kernel is the covariance of the output of a second-order system. In contrast with the previous examples, however, we consider a particular system and analyze it by using numerical methods. Obtaining analytic results for systems whose dimension is greater than one is straightforward, but *extremely* tedious. Let us assume that the kernel $\mathbf{K_y}(t, \tau)$ is the covariance function given by Equation 2.27 and illustrated by Figure 2.29. The state equations for generating $y(t)$ are specified by Equation 2.28. In addition, let us set $T_f = 2$ and $T = 0$.

First, we need the matrix $\mathbf{W}(\lambda)$. By substituting Equation 2.28 into 3.14 we obtain

$$\mathbf{W}(\lambda) = \begin{bmatrix} 0 & 1 & 0 & 0 \\ -10 & -2 & 0 & 160 \\ -\dfrac{1}{\lambda} & 0 & 0 & 10 \\ 0 & 0 & -1 & 2 \end{bmatrix}. \tag{3.69}$$

In order to determine $A(\lambda)$ for a given λ we need to find the transition matrix of $\mathbf{W}(\lambda)$ at $T = 2$; i.e., we need to determine the exponential matrix function of $\mathbf{W}(\lambda)T$. We calculated the results for this example by summing the first thirty terms in the series representation in a nested fashion. We would suggest, however, using the Faddeeva algorithm in any future calculation.[43]

Once we find this transition matrix, we use 3.22b to determine $\mathbf{\Phi}_\eta(T_f, T_0 : \lambda)$, and then we compute the determinant of this matrix for $A(\lambda)$. By varying the parameter λ and repeating this procedure, we can plot $A(\lambda)$ versus λ, possibly using a locally iterative technique to locate the eigenvalues, the roots of $A(\lambda) = 0$. The resulting curve is presented in Figure 3.1. This type of curve is typical of those that we observe for the function $A(\lambda)$ with the zero crossing being the eigenvalues. For large values of λ, we approach the constant indicated by the Fredholm determinant

$$\exp\left(-\int_{T_0}^{T_f} \mathrm{Tr}[\mathbf{F}(t)]dt\right) = e^4 \tag{3.70}$$

As we decrease λ, we find that in the region corresponding to the $2WT + 1$ most significant eigenvalues, for this example 3, the function is oscillating nonperiodically in a well-behaved manner. However, as we let λ approach the region of the less significant eigenvalues, this

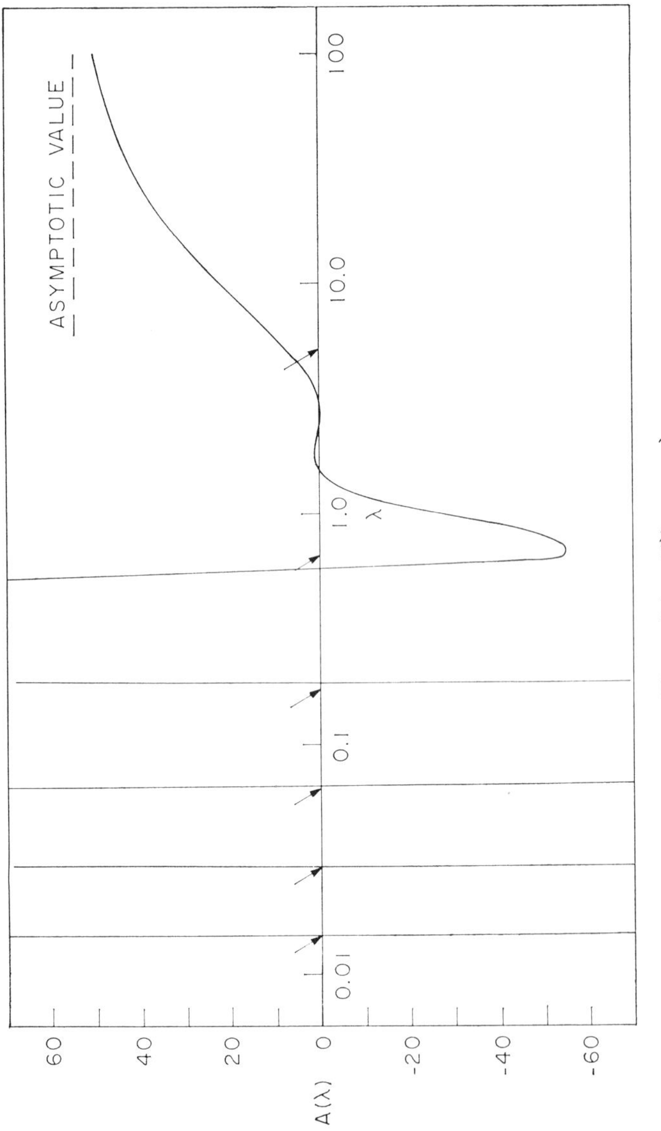

Figure 3.1 $A(\lambda)$ versus λ.

oscillation becomes extremely large. Eventually, the numbers become so large that it becomes difficult to compute them effectively.

When this situation occurs, it is probably better to determine the eigenvalues by using an asymptotic technique as suggested by Capon.*[16] We have indicated the location of the eigenvalues as given by this technique by the arrowheads in Figure 3.1. As the eigenvalues decrease, the agreement becomes quite good. We should point out, however, the pronounced difference that appears when we try to extend the technique in order to evaluate the significant eigenvalues. Since this state variable technique is useful for computing the significant eigenvalues, one could combine this method with an asymptotic method such as this to find all the eigenvalues accurately and conveniently. The eigenvalue behavior for this example is summarized in Figure 3.2, where we have plotted the first most significant eigenvalues versus the interval length $T_f - T_0 = T$. We see that the curves increase monotonically with T. This is in agreement with the formula[33]

$$\frac{\partial \lambda_i(t)}{\partial T} = \lambda_i(t)\,|\phi_i(T)|^2, \tag{3.71}$$

We also note that the number of significant eigenvalues increases with T, reflecting the $2WT + 1$ measure. Finally we note the crossing implying the existence of multiple-order eigenvalues.† One can attribute this to the spectral peaks being displaced from the origin.

The calculation of the Fredholm determinant, or the related function $d(\lambda)$ as given by Equation 3.47 is done straightforwardly by evaluating $A(-\lambda)$ and the multiplicative coefficients. Using the same algorithm as for the eigenvalue determination, we have plotted in Figure 3.3 the function $\ln[d(\lambda)]$ versus λ for time interval lengths of $\frac{1}{2}$, 1, 2, and 4. We compare our results with those derived using Equation 3.48, where we assume that the time interval is large compared to the inverse of the process bandwidth. We can see that for $T \geqq 2$ we are very close to the asymptotic results indicated by this stationary process, large-time-interval analysis.

* This technique approximates the eigenvalues by evaluating the spectrum at $n\pi/T$ or $(n + \frac{1}{2})\pi/T$ depending upon whether the difference between the denominator and numerator polynomials is twice an odd or even number, respectively. The technique is asymptotically valid as n increases.

† As noted earlier, the dimension of the system generating the kernel sets an upper bound on the eigenvalue multiplicity since the rank of $\boldsymbol{\phi\eta}(T_f, T_0 : \lambda)$ is between zero and this dimension.

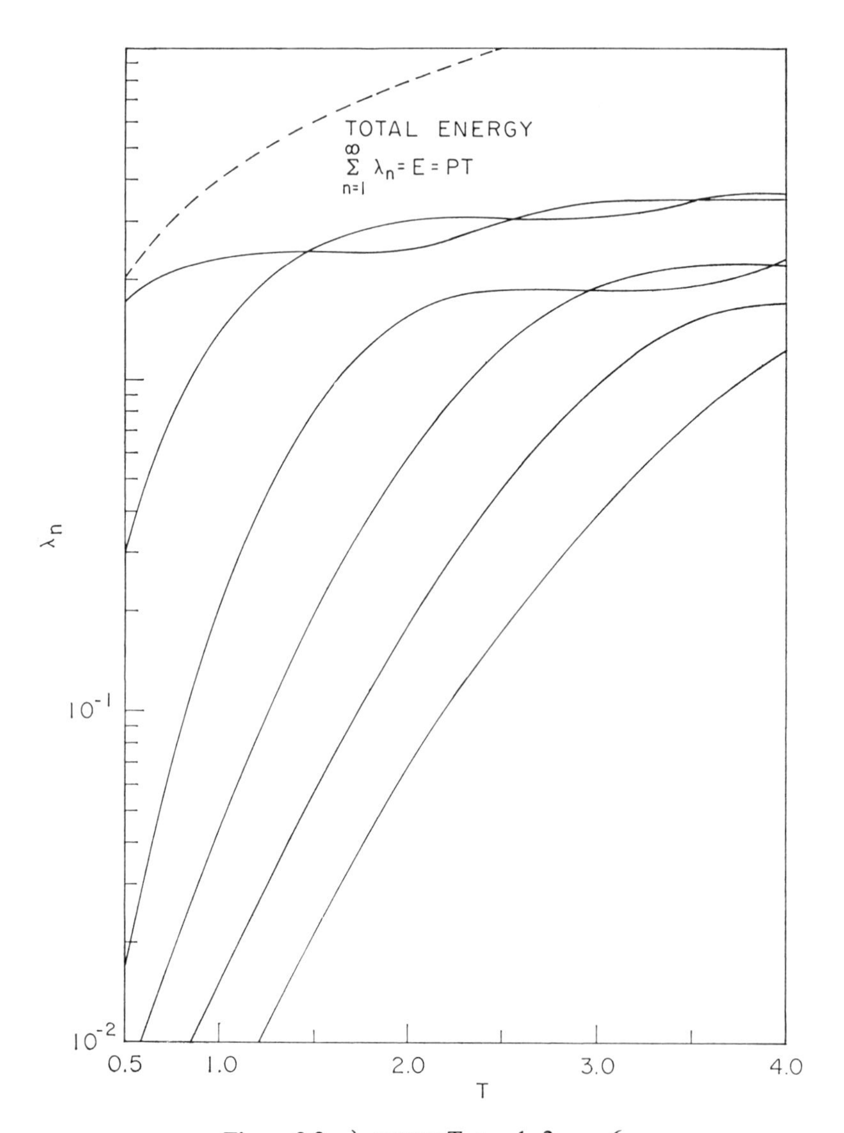

Figure 3.2 λ_n versus T, $n = 1, 2, \ldots, 6$.

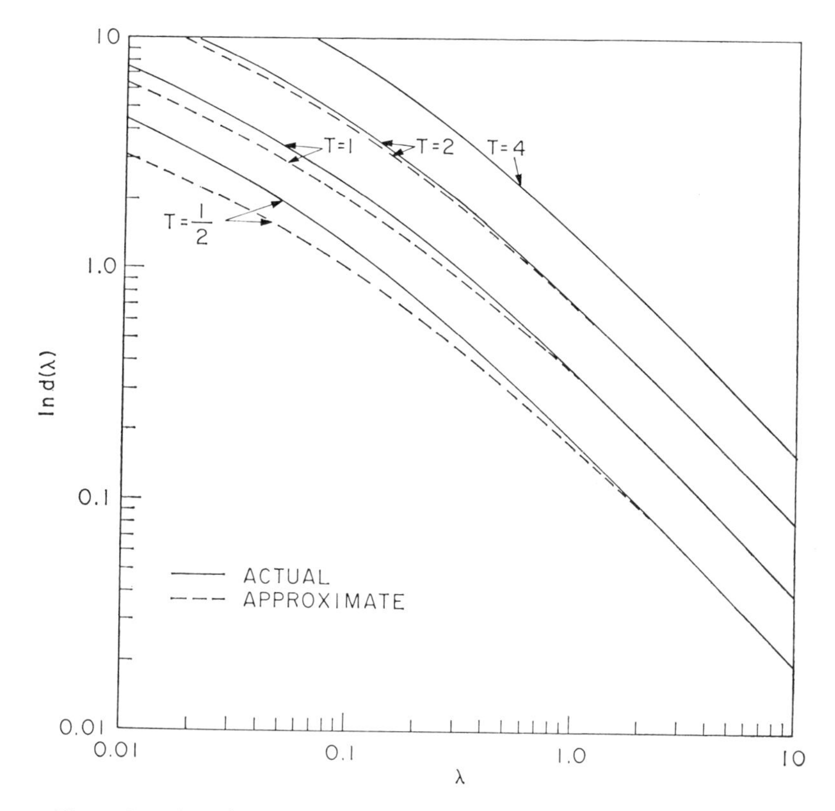

Figure 3.3 $\ln[d(\lambda)]$ versus λ for the second-order process of Figure 2.2.

Example 3.4 The Higher-Order Butterworth Processes*

We can analyze the higher-order Butterworth processes just as we did the previous example. Since these processes are often employed as a theoretical tool, it is useful to summarize the results of such an analysis.

The spectrum of the nth-order Butterworth process with power P is given by

$$S_y(\omega : n) = \frac{2nP}{k} \frac{\sin(\pi/2n)}{1 + (\omega/k)^{2n}}.$$

* The second-order Butterworth process has been considered by Youla.[73] The complexity of his analytic results indicate the appropriateness of a numerical algorithm.

The results of an analysis such as the one done in the previous section are given in Figures 3.4 and 3.5. The state matrices are specified on the respective figures. We have also indicated the limiting situation specified by an ideal band-limited process resulting in the prolate spheroidal wavefunctions in Figures 3.6 and 3.7.[58,59] We see that for $n = 4$ we have substantially reached the ideal band-limited case for all practical purposes.*

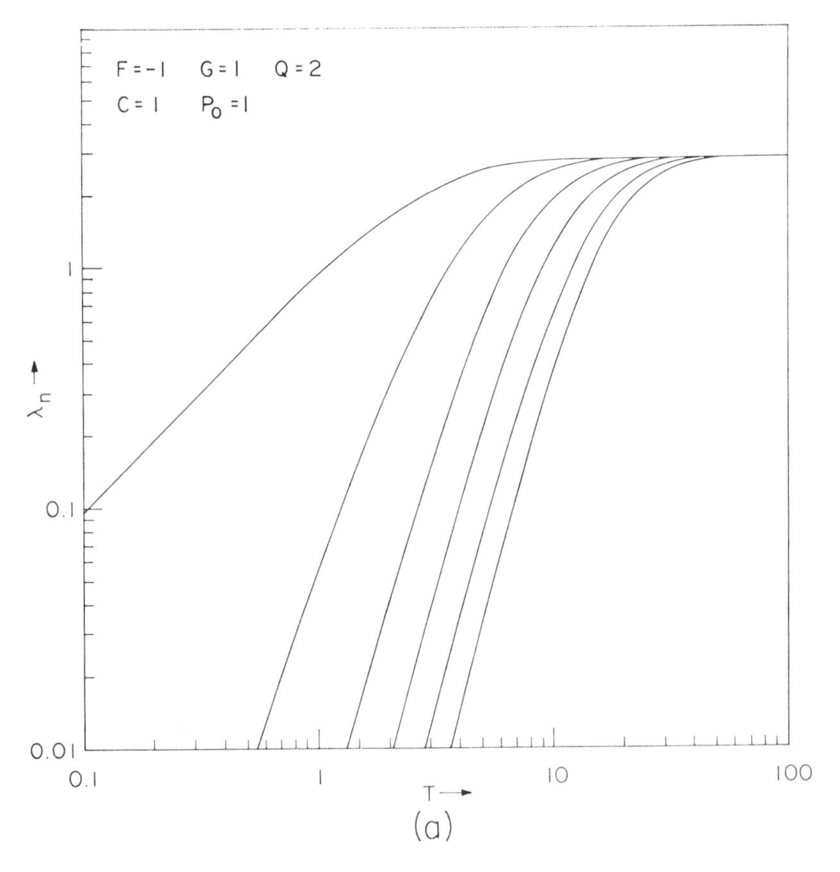

Figure 3.4 (a) λ_n versus T, $n = 1, 2, \ldots, 6$, for a first-order Butterworth process. (b) λ_n versus T, $n = 1, 2, \ldots, 6$, for a second-order Butterworth process. (c) λ_n versus T, $n = 1, 2, \ldots, 6$, for a third-order Butterworth process. (d) λ_n versus T, $n = 1, 2, \ldots, 6$, for a fourth-order Butterworth process.

* The data of Ref. 62 only allowed evaluation for T less than 16.

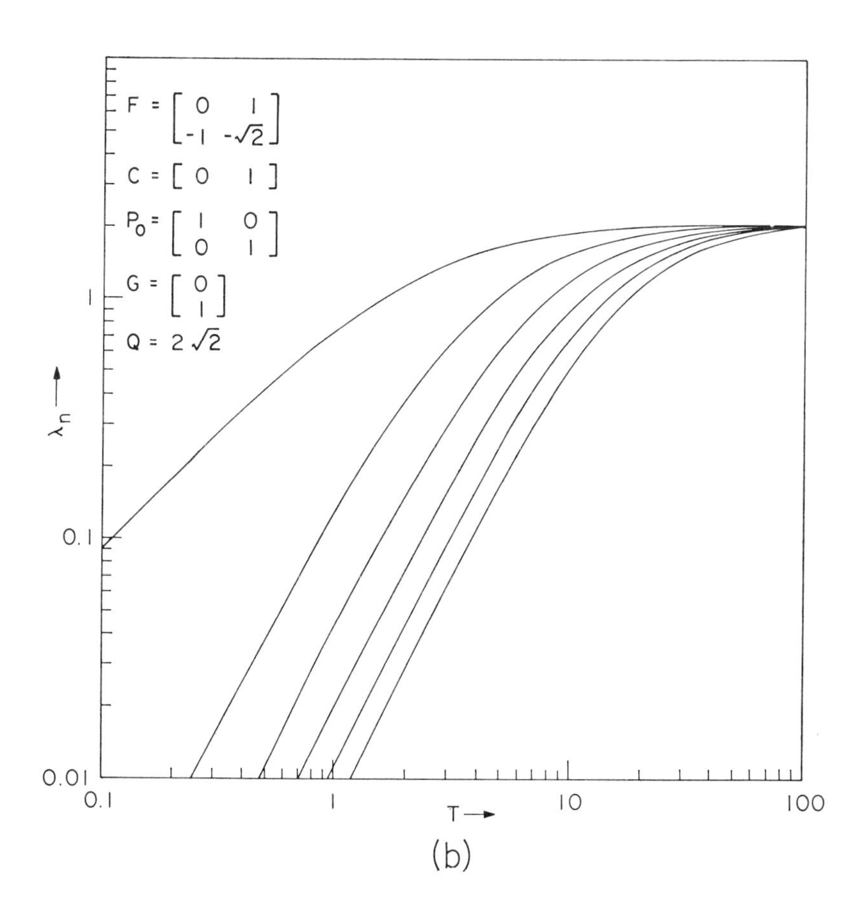

(b)

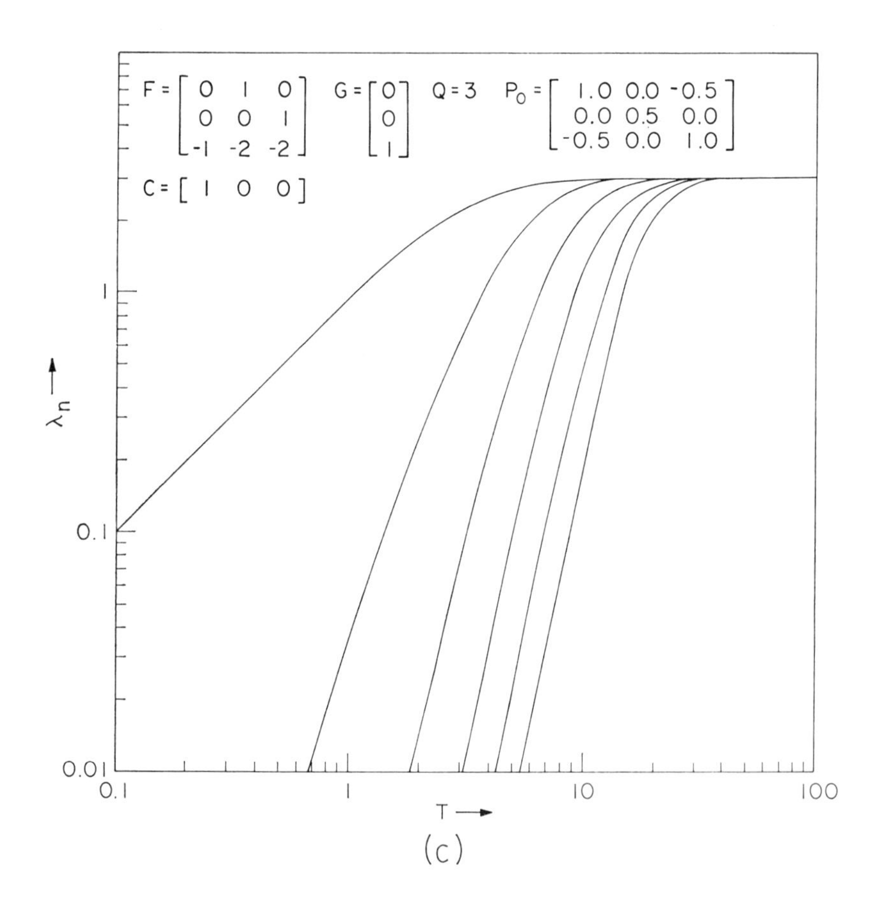
F = [0 1 0 ; 0 0 1 ; -1 -2 -2]
G = [0 ; 0 ; 1]
Q = 3
P0 = [1.0 0.0 -0.5 ; 0.0 0.5 0.0 ; -0.5 0.0 1.0]
C = [1 0 0]
λn
1
0.1
0.01
0.1
1
10
100
T

(c)

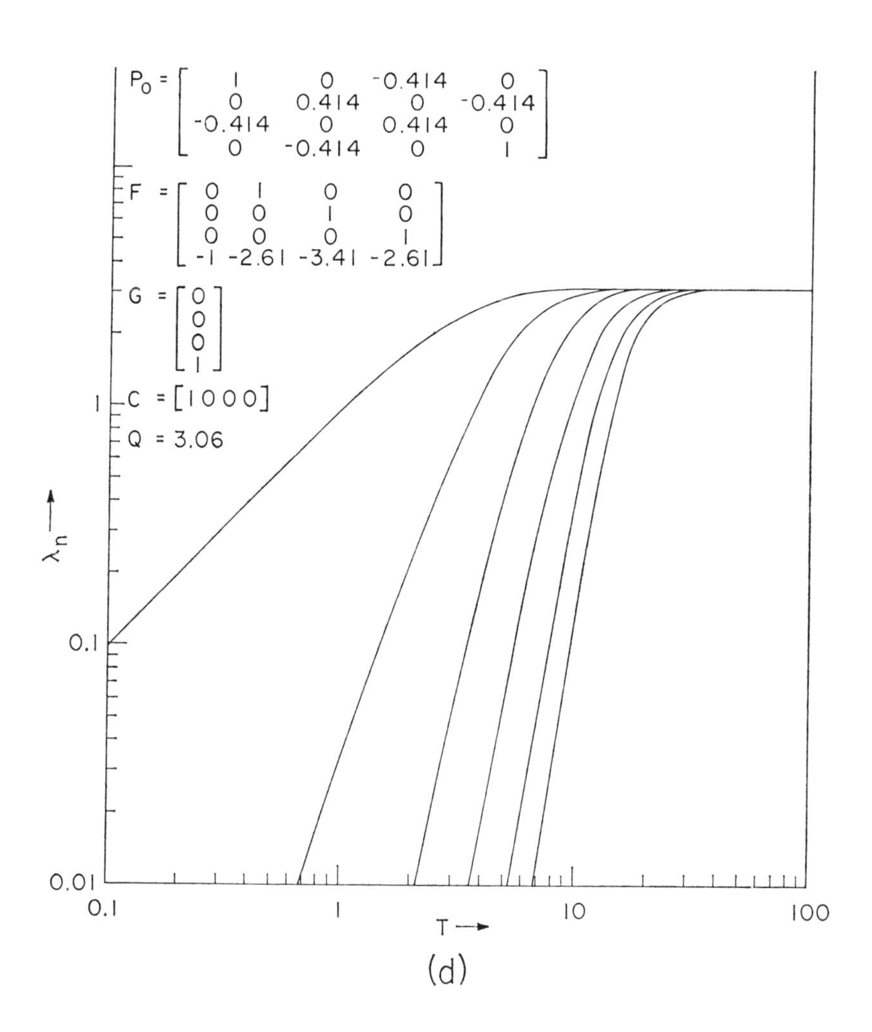
P0 = [1 0 -0.414 0
0 0.414 0 -0.414
-0.414 0 0.414 0
0 -0.414 0 1]
F = [0 1 0 0
0 0 1 0
0 0 0 1
-1 -2.61 -3.41 -2.61]
G = [0
0
0
1]
C = [1 0 0 0]
Q = 3.06
λn
1
0.1
0.01
0.1
1
10
100
T
(d)

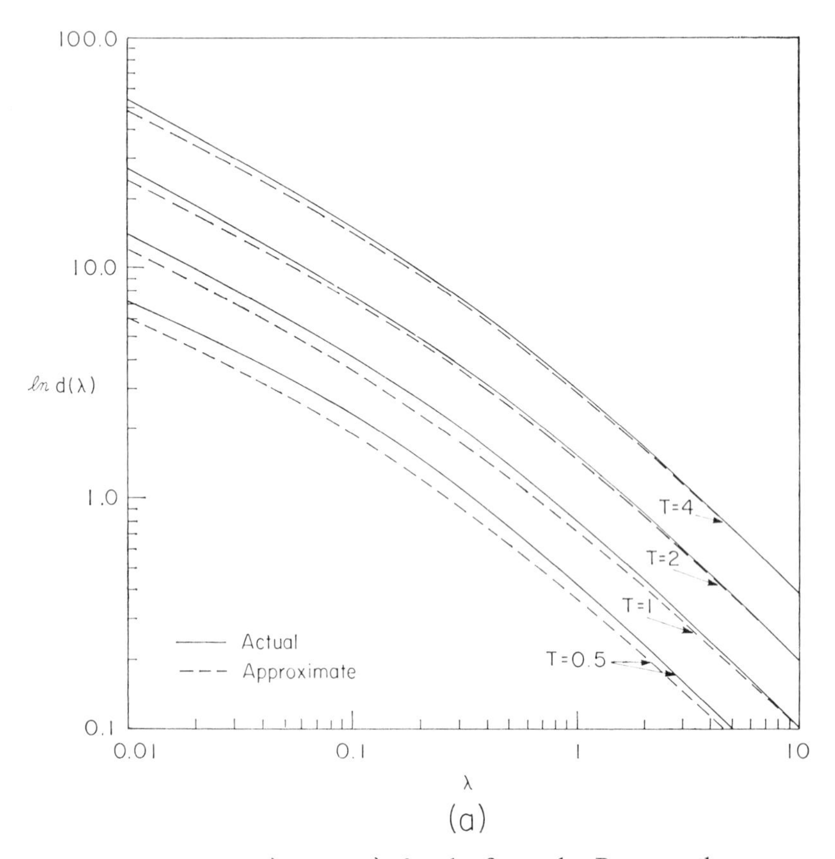

Figure 3.5 (a) ln $d(\lambda)$ versus λ, for the first-order Butterworth process. (b) ln $d(\lambda)$ versus λ for a second-order Butterworth process. (c) ln $d(\lambda)$ versus λ for the third-order Butterworth process. (d) ln $d(\lambda)$ versus λ for the fourth-order Butterworth process.

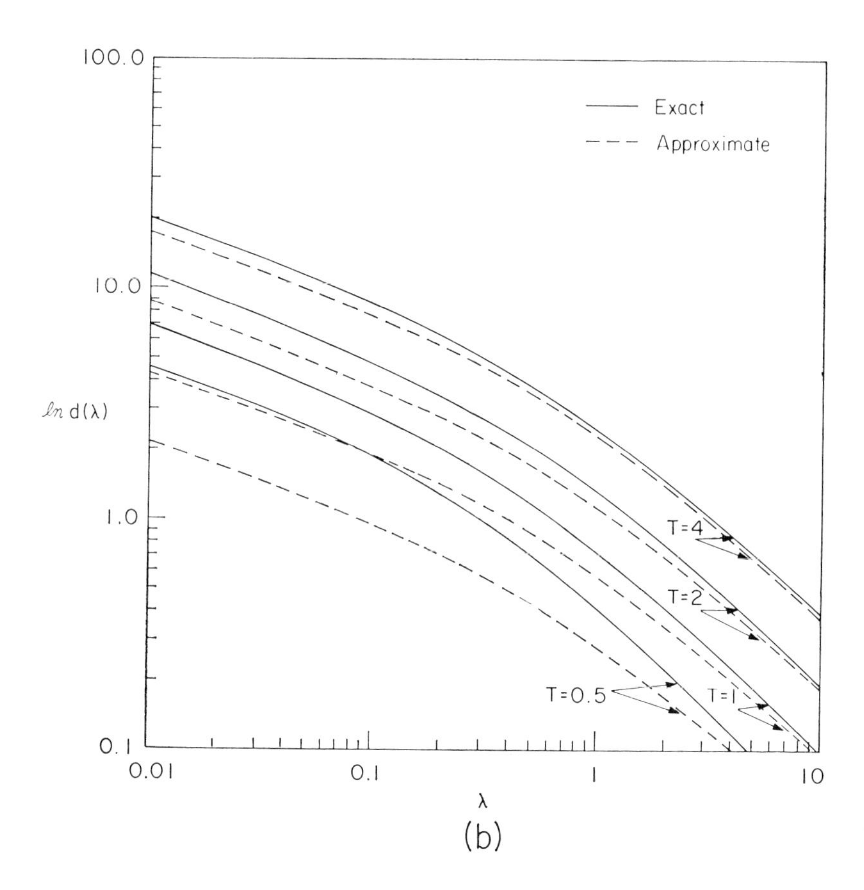
100.0
10.0
1.0
0.1
0.01
0.1
1
10
Exact
Approximate
ℓn d(λ)
λ
T=4
T=2
T=0.5
T=1

(b)

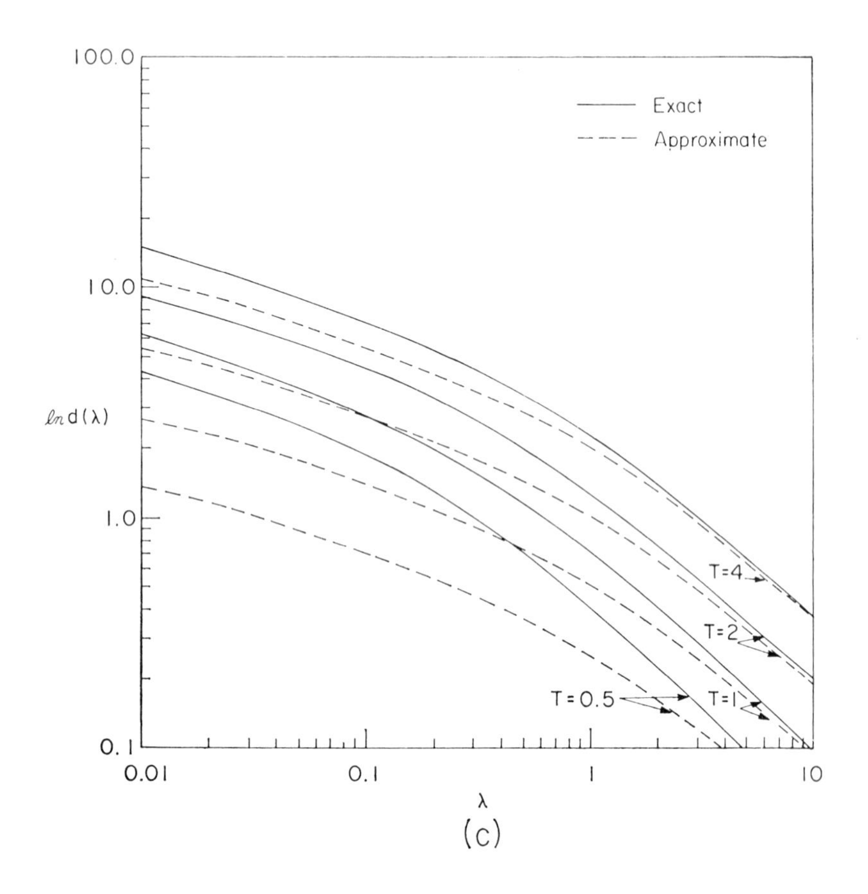

100.0
10.0
1.0
0.1
0.01
0.1
1
10
$\ell n\, d(\lambda)$
λ
Exact
Approximate
T=4
T=2
T=1
T=0.5

(c)

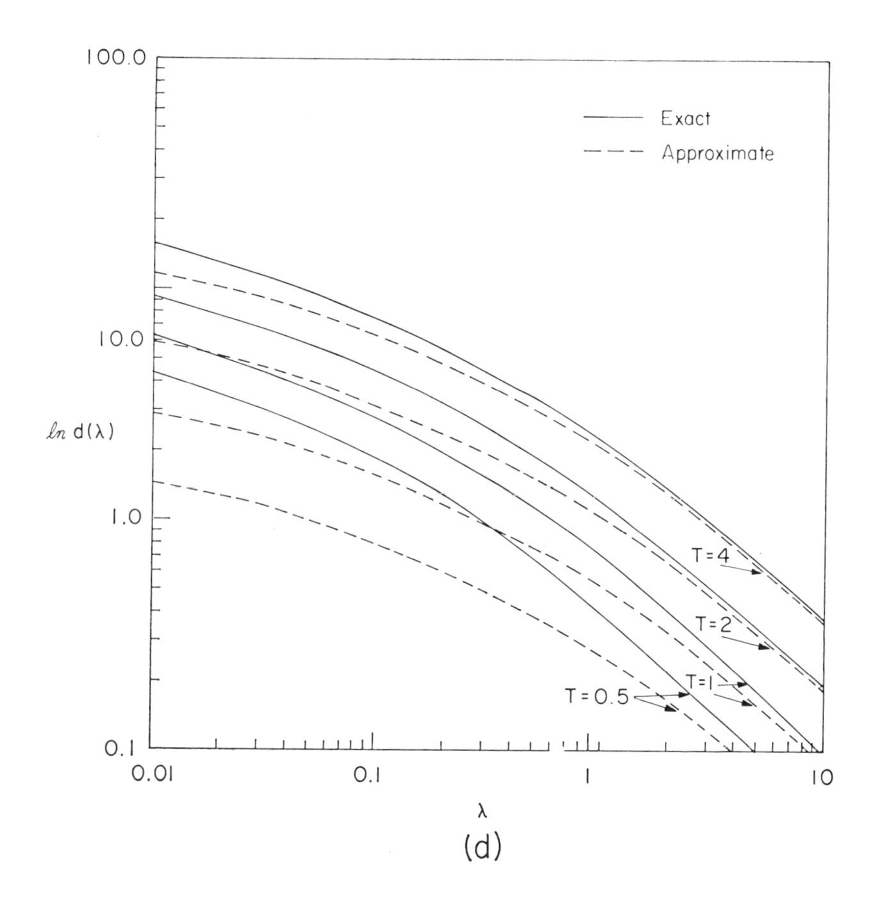

(d)

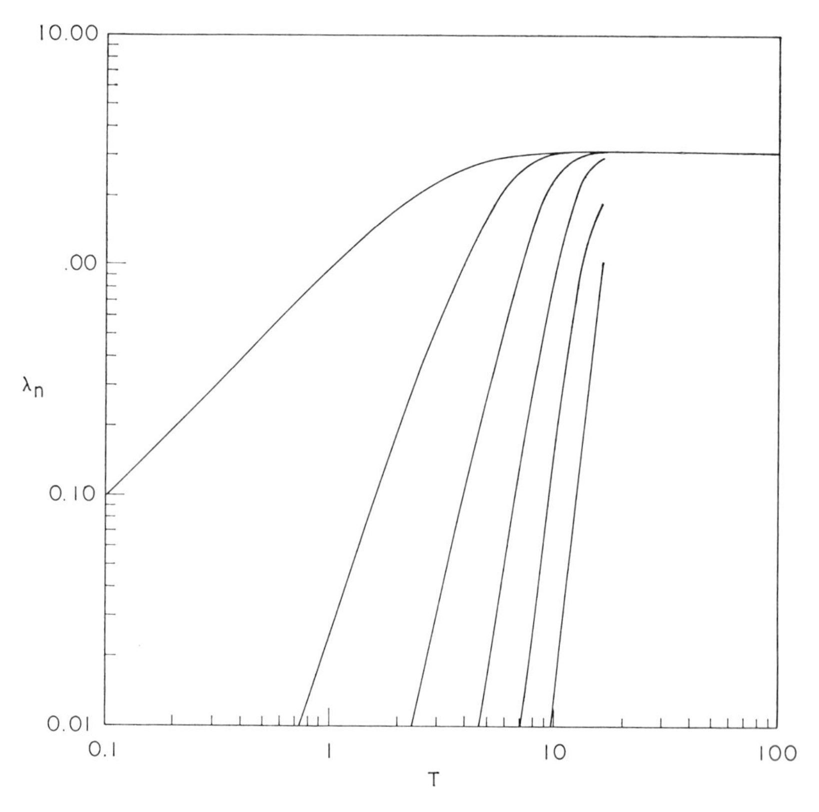

Figure 3.6 λ_n versus T, $n = 1, 2, \ldots, 6$, for ideal band-limited process.

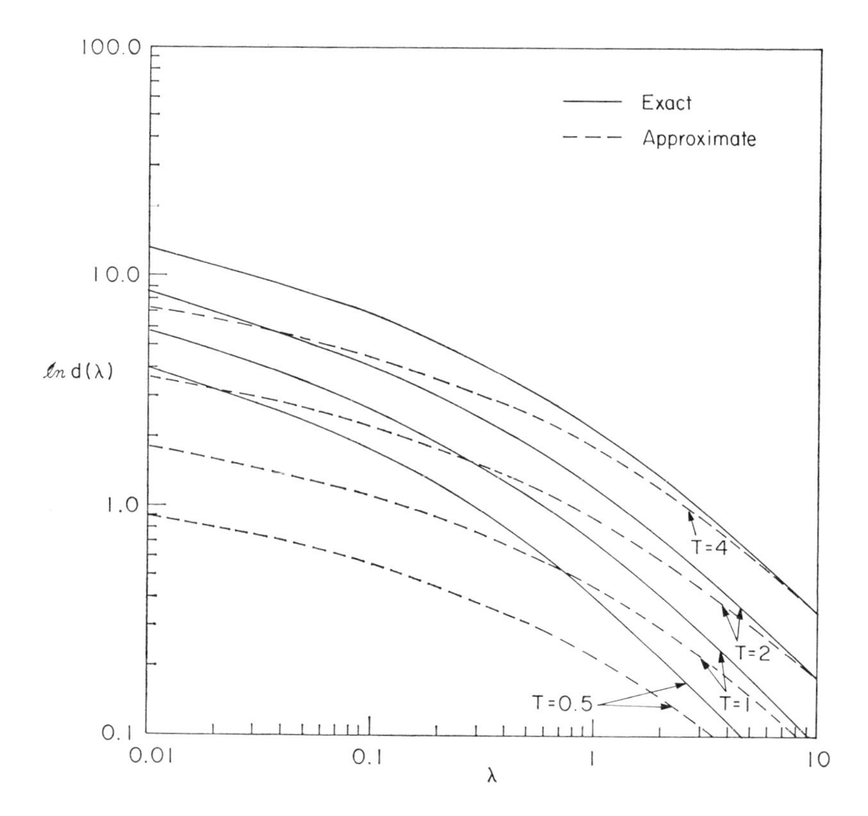

Figure 3.7 ln $d(\lambda)$ versus λ for an ideal band-limited process.

We have indicated the results that a numerical analysis typically yields. We used one particular kernel in the example for both of the integral equations. We can consider several other types of kernels. We also can analyze a large class of time-varying kernels. For those which are covariances of processes that are the output of a constant-parameter system, the analysis is identical. For those with time-varying parameters, the most significant difference is that we do not have the luxury of using the matrix exponential for many of our calculations.

3.4 Discussion of Results for Homogeneous Fredholm Integral Equations

In this chapter we have formulated a state variable approach for solving homogeneous Fredholm integral equations. As we indicated earlier, the technique has several advantages, particularly from a computational viewpoint. Consequently, in problems where we need to evaluate the eigenvalues and/or eigenfunctions directly, we have a very general method available which allows us to make just such an evaluation with a minimum effort.

Quite often we do not need these solutions directly, but we require an expression involving them. In many cases, we should be able to determine such expressions in a closed form by using the theory that we have developed, for example, as we did with the Fredholm determinant function.

Proceeding with our discussion, several comments are in order regarding the desirability of having a convenient method of evaluating this Fredholm determinant function.

1. In the problem of detecting Gaussian signals in Gaussian noise, the Fredholm determinant function enters in two ways. First, it appears as the bias in calculating the threshold of the likelihood receiver. More importantly, it is intimately involved in the calculation of performance bounds for this problem.

2. In an estimation theory context, if we consider the problem of estimating the parameters of a Gaussian process (or the system identification problem), we can show that the Cramer-Rao and Bhattacharyya bound can be determined using this function.

Finally, the solution method we have developed is important in itself. As we shall see in Chapter 5, the concept of requiring a determinant to vanish for the existence of a solution is important. In this chapter, we find a similar set of homogeneous differential equations and boundary conditions which specifies the solution to the optimal signal design problem for colored noise problems. The method of solving these equations is exactly analogous to the eigenvalue problem in that it requires the vanishing of a determinant for a solution to exist.

Now that we have studied the solution of homogeneous Fredholm equations, let us examine the inhomogeneous equation.

4 Inhomogeneous Fredholm Integral Equations

Inhomogeneous Fredholm integral equations are of considerable interest in communication theory. In this chapter we use the results of Chapter 2 to develop a state variable method of solving these equations. As before, there are some significant advantages to the methods introduced.*

One of the more important applications of this integral equation is in determining the structure of optimal receivers when there is colored noise present in the observed signal in addition to the signal component. Since we study several aspects of this problem in detail both in this chapter and the next, we first pause briefly to review the communication models that lead to this equation. This does not imply that this problem is the only place where can we apply our methods. Other possible applications include the solution of finite-time Wiener-Hopf equations in order to find optimal estimators for processes, as we do in Chapter 6, and the calculation of the bounds on the estimation accuracy of signal parameters.

After this brief discussion we present a derivation that reduces the integral equation to a pair of vector differential equations and associate boundary conditions. The results follow quickly since we have derived the requisite formulas in Chapter 2. Since these equations have appeared in the literature in another related context, namely the linear smoother, we have several solution methods available.[15,49,28] We devote a section

* Several of the references cited at the beginning of Chap. 3 also discuss solving the inhomogeneous equations.

to introducing these methods and to making comments upon their applicability.

Finally, we consider three examples. Two of these are worked analytically, and we present a numerical approach for the third.

4.1 Communication Models for Detecting Signals in Colored Noise

We introduce the application of the inhomogeneous Fredholm integral equation in the context of detecting a known signal in additive colored noise. This is the simplest problem that leads to the use of this equation, and it is a useful vehicle for conveying several of the issues that we discuss. After this introduction we discuss where it appears in several other applications.

Figure 4.1 is a model of a communication system with an additive colored noise channel. We have a transmitter which on hypothesis 1 transmits $\mathbf{s}_1(t)$; on hypothesis 0, it transmits $\mathbf{s}_0(t)$ over the time interval $T_0 \leqq t \leqq T_f$. For discussion purposes we assume that $\mathbf{s}_1(t)$ is $\mathbf{s}(t)$ while $\mathbf{s}_0(t)$ is zero over $[T_0, T_f]$. The channel adds a vector colored Gaussian noise process to the signal. We assume that this colored noise consists of two independent components. The first component is a random process $\mathbf{y}(t)$ that is generated according to the methods we discussed in Chapter 2. The second component is an independent white Gaussian process $\mathbf{w}(t)$ that has a covariance

$$E[\mathbf{w}(t)\mathbf{w}^T(\tau)] = \mathbf{R}(t)\,\delta(t-\tau). \tag{4.1}$$

Consequently, we have the following detection problem:

$$\begin{aligned} &\text{on } H_1\ \mathbf{r}(t) = \mathbf{s}(t) + \mathbf{y}(t) + \mathbf{w}(t), \qquad && T_0 \leqq t \leqq T_f, \\ &\text{on } H_0\ \mathbf{r}(t) = \mathbf{y}(t) + \mathbf{w}(t), && T_0 \leqq t \leqq T_f. \end{aligned} \tag{4.2}$$

Note that we do not preclude the situation where $\mathbf{y}(t)$ depends upon the transmitted signal $\mathbf{s}(t)$. In this case, some of the coefficient matrices in the state representation discussed in Chapter 2 depend upon $\mathbf{s}(t)$, and, as a result, would in general lead to a time-varying analysis.

Under these assumptions it is straightforward to show that the optimal receiver can be realized as indicated in Figure 4.1. This realization is a correlation receiver. We multiply (dot product) the received signal $\mathbf{r}(t)$ with a function $\mathbf{g}(t)$, and then integrate over the observation interval; i.e., the sufficient statistic for the decision device is

$$l(\mathbf{r}) = \int_{T_0}^{T_f} \mathbf{r}^T(\tau)\mathbf{g}(\tau)\,d\tau. \tag{4.3}$$

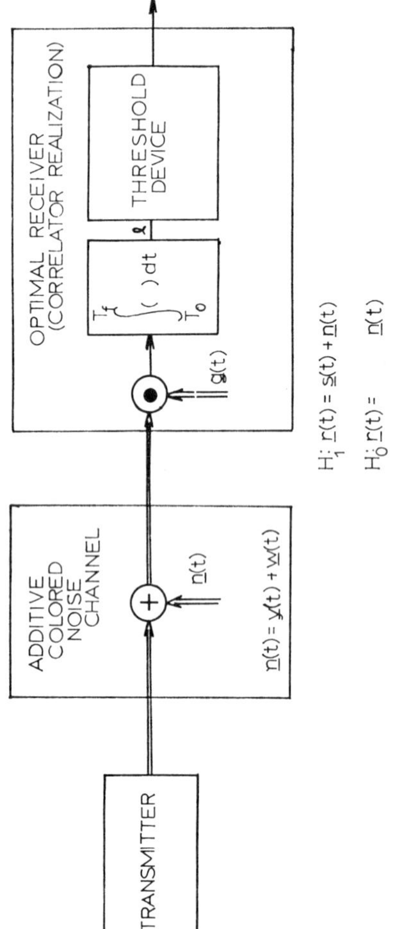

Figure 4.1 System model for the detection of a known signal in colored noise.

The correlating signal is the solution to an inhomogeneous Fredholm integral equation.

This integral equation has the form

$$\int_{T_0}^{T_f} \mathbf{K}_{\mathbf{y}}(t, \tau)\mathbf{g}(\tau)\, d\tau + \mathbf{R}(t)\mathbf{g}(t) = \mathbf{s}(t), \qquad T_0 \leqq t \leqq T_f, \tag{4.4}$$

where $\mathbf{K}_{\mathbf{y}}(t, \tau)$, the kernel of the equation, is the covariance of the random process $\mathbf{y}(t)$, which we assumed is generated according to the methods discussed in Chapter 2; $\mathbf{s}(t)$ is a known vector function, the transmitted signal; $\mathbf{R}(t)$ is the covariance weighting matrix of $\mathbf{w}(t)$, which is assumed to be positive definite; and $\mathbf{g}(t)$ is the desired solution, the correlating signal.

The positive definiteness of $\mathbf{R}(t)$ implies that each component of the observed signal contains a white noise that cannot be eliminated. This implies that Equation 4.4 is a Fredholm integral equation of the second kind; consequently, a unique bounded solution exists. This also precludes Fredholm equations of the first kind. Equations of this type are often associated with singular detection problems, and their solutions usually contain singularity functions at the end points of the observation interval.*

One can also find the performance measure for this system. It is again straightforward to show that this measure, usually termed d^2, is given by

$$d^2 = \int_{T_0}^{T_f} \mathbf{s}^T(\tau)\mathbf{g}(\tau)\, d\tau, \tag{4.5}$$

where $\mathbf{s}(t)$ and $\mathbf{g}(t)$ are defined above. Error probabilities, detection, and false alarm probabilities can all be determined in terms of this measure; for example,

$$P_D = \operatorname{erfc}_*\left(\frac{\ln \eta}{d} - \frac{d}{2}\right), \tag{4.6a}$$

$$P_F = \operatorname{erfc}_*\left(\frac{\ln \eta}{d} + \frac{d}{2}\right), \tag{4.6b}$$

where

$$\operatorname{erfc}_*(x) = \int_x^\infty \frac{1}{\sqrt{2\pi}}\, e^{-X^2/2}\, dX. \tag{4.6c}$$

* Ref. 22, pp. 240–242; Ref. 50.

We encounter this integral equation in several other applications. In many of these, one uses a model for the detection problem which is given by*

$$\text{on } H_1: \quad \tilde{r}(t) = \tilde{b}\tilde{s}(t) + \tilde{y}(t) + \tilde{w}(t), \tag{4.7}$$

$$\text{on } H_0: \quad \tilde{r}(t) = \tilde{y}(t) + \tilde{w}(t),$$

where $\tilde{b}$ is a complex zero mean, Gaussian random variable with variance $2\sigma_b^2$,

$$E[bb^*] = 2\sigma_b^2. \tag{4.8}$$

This model is appropriate for many problems where the hypothesis does not affect the statistics of the noise $\mathbf{y}(t)$. This would be the case in some communication channels. If one models the clutter, or reverberation, as being generated by a state variable representation with signal-dependent coefficient matrices, then an important application is the associated detection problem as stated by Equation 4.7.

The optimum receiver for this problem simply squares the magnitude of the output of the correlator in Figure 4.1 before comparing it with a threshold. The associate performance measure is given by

$$P_D = [\mathbf{P}_F]^{1/(1+2\sigma_b{}^2 d^2)}, \tag{4.9}$$

where d^2 is as specified by Equation 4.5.

If the noise is hypothesis dependent, we are led to an appreciably more difficult problem. Problems such as this are discussed in detail in Reference 68. In such problems one is faced with solving Wiener-Hopf integral equations, and our results in this chapter can be, and are, used in Chapter 6 to solve this important integral equation in communication theory. There are several other applications of the inhomogeneous equation in communication theory, for it arises quite naturally in problems involving colored noise processes. As we mentioned earlier, we study the inhomogeneous integral Equation 4.4 in the context of the additive noise detection problem. We emphasize however, that the techniques developed are general in that they do not need to be considered in this particular context. We now develop our solution method.

* The tilde implies that the variables are complex. This arises when one uses complex notation to represent bandpass processes and functions. We discuss this issue for state variable processes in Appendix B.

4.2 The State Variable Solution to Inhomogeneous Fredholm Integral Equations

The integral equation that we wish to solve is given by Equation 4.4. As mentioned earlier, in contrast to the homogeneous equation, we can find a unique solution when $\mathbf{R}(t)$ is positive definite. We assume that this condition is satisfied. Let us rewrite Equation 4.4 in the form*

$$\mathbf{g}(t) = \mathbf{R}^{-1}(t)\left[\mathbf{s}(t) - \mathbf{C}(t)\left(\int_{T_0}^{T_f} \mathbf{K_x}(t, \tau)\mathbf{C}^T(\tau)\mathbf{g}(\tau)\, d\tau\right)\right], \quad T_0 \leqq t \leqq T_f, \quad (4.10)$$

where we have substituted Equation 2.6 for $\mathbf{K_y}(t, \tau)$. If in 4.10 we set

$$\mathbf{g}(\tau) = \mathbf{f}(\tau), \quad (4.11)$$

we have the result that the integral enclosed in parentheses is the function $\xi(t)$ as defined by Equation 2.37. Consequently, for the inhomogeneous equation we define $\xi(t)$ to be

$$\xi(t) = \int_{T_0}^{T_f} \mathbf{K_x}(t, \tau)\mathbf{C}^T(\tau)\mathbf{g}(\tau)\, d\tau, \qquad T_0 \leqq t \leqq T_f, \quad (4.12)$$

so that Equation 4.10 becomes

$$\mathbf{g}(t) = \mathbf{R}^{-1}(t)[\mathbf{s}(t) - \mathbf{C}(t)\xi(t)], \qquad T_0 \leqq t \leqq T_f. \quad (4.13)$$

For the class of kernels that we are considering, we have shown in Section 2.3, Equations 2.40, and 2.42, that the functional defined by Equation 4.12 may be represented as the solution to the following differential equations:

$$\frac{d\xi(t)}{dt} = \mathbf{F}(t)\xi(t) + \mathbf{G}(t)\mathbf{Q}\mathbf{G}^T(t)\eta(t), \quad T_0 \leqq t \leqq T_f, \quad (4.14)$$

$$\frac{d\eta(t)}{dt} = -\mathbf{C}^T(t)\mathbf{g}(t) - \mathbf{F}^T(t)\eta(t), \qquad T_0 \leqq t \leqq T_f, \quad (4.15)$$

plus a set of boundary conditions. If we substitute Equation 4.13 for $\mathbf{g}(t)$, Equation 4.15 becomes

$$\frac{d\eta(t)}{dt} = \mathbf{C}^T(t)\mathbf{R}^{-1}(t)\mathbf{C}(t)\xi(t) - \mathbf{F}^T(t)\eta(t) - \mathbf{C}^T(t)\mathbf{R}^{-1}(t)\mathbf{s}(t), \quad T_0 \leqq t \leqq T_f. \quad (4.16)$$

* The first part of this approach was suggested by Collins.[17]

If we write Equations 4.14 and 4.16 in an augmented vector form, we consequently have shown that the inhomogeneous integral equation can be reduced to the differential equations

$$\frac{d}{dt}\begin{bmatrix}\boldsymbol{\xi}(t)\\ \boldsymbol{\eta}(t)\end{bmatrix} = \begin{bmatrix}\mathbf{F}(t) & \mathbf{G}(t)\mathbf{Q}\mathbf{G}^T(t)\\ \mathbf{C}^T(t)\mathbf{R}^{-1}(t)\mathbf{C}(t) & -\mathbf{F}^T(t)\end{bmatrix}\begin{bmatrix}\boldsymbol{\xi}(t)\\ \boldsymbol{\eta}(t)\end{bmatrix} - \begin{bmatrix}\mathbf{0}\\ \mathbf{C}^T(t)\mathbf{R}^{-1}(t)\mathbf{s}(t)\end{bmatrix}, \quad T_0 \leqq t \leqq T_f. \tag{4.17}$$

where the boundary conditions follow from Equations 2.45 and 2.43,

$$\boldsymbol{\xi}(T_0) = \mathbf{P}_0\,\boldsymbol{\eta}(T_0), \tag{4.18}$$

$$\boldsymbol{\eta}(T_f) = \mathbf{0}. \tag{4.19}$$

The solution $\mathbf{g}(t)$ of the original integral equation is related to the differential equation solution by Equation 4.13. As a result, we have reduced the problem of solving the inhomogeneous integral equation to one of solving a pair of vector differential equations with mixed or split boundary conditions.

We note that the coefficient matrix in Equation 4.17 is similar to that which appeared in the homogeneous equation. If we replace the positive definite matrix $\mathbf{R}(t)$ with $-\lambda$ we would have the same matrix. We note from our discussion of the Fredholm determinant that one wants to compute the transition matrix of this coefficient matrix to find $D_{\mathscr{F}}(z)$.

In the context of optimum receiver performance, Equation 4.13 leads to an appealing interpretation. Substituting Equation 4.13 in Equation 4.5, we obtain

$$d^2 = \int_{T_0}^{T_f} \mathbf{s}^T(t)\mathbf{g}(t)\,dt = \int_{T_0}^{T_f} \mathbf{s}^T(t)\mathbf{R}^{-1}(t)\mathbf{s}(t)\,dt - \int_{T_0}^{T_f} \mathbf{s}^T(t)\mathbf{R}^{-1}(t)\mathbf{C}(t)\boldsymbol{\xi}(t)\,dt. \tag{4.20}$$

The first term is simply the pure white noise performance, d_w^2. The second term represents the degradation d_g^2, caused by the presence of the colored noise in the observation. We have

$$d^2 = d_w^2 - d_g^2, \tag{4.21a}$$

where

$$d_w^2 = \int_{T_0}^{T_f} \mathbf{s}^T(t)\mathbf{R}^{-1}(t)\mathbf{s}(t)\,dt, \tag{4.21b}$$

$$d_g^2 = \int_{T_0}^{T_f} \mathbf{s}^T(t)\mathbf{R}^{-1}(t)\mathbf{C}(t)\boldsymbol{\xi}(t)\,dt. \tag{4.21c}$$

4.3 Methods of Solving the Differential Equations for the Inhomogeneous Fredholm Equation

In the last section we derived a pair of vector differential equations that implicitly specified the solution of Equation 4.4. As we shall see in Chapter 6, these equations appear in the optimal smoother, or the state variable formulation of the unrealizable filter.* In this section we exploit the methods that have been developed in the literature for solving the smoothing equations in order to solve the differential equations for the inhomogeneous equation. Since these methods have evolved from the smoothing theory literature, the material in this section draws heavily upon References 15, 49, and 28.

Let us outline our approach. We develop three methods, each of which has a particular application. The first method is useful for obtaining analytic solutions. The second is an intermediate result in the development of the third method. It is useful because it introduces some important concepts and results. The third method is applicable to finding numerical solutions since the first two methods have some undesirable aspects. We should also note that the methods build upon one another. Consequently, one needs to read the entire section to understand the development of the last method.

Before proceeding, let us summarize the results that we need from the previous section and introduce some notation that we require. Rewriting Equation 4.17, we want to solve the differential equations

$$\frac{d}{dt}\begin{bmatrix}\boldsymbol{\xi}(t)\\ \boldsymbol{\eta}(t)\end{bmatrix} = \mathbf{W}(t)\begin{bmatrix}\boldsymbol{\xi}(t)\\ \boldsymbol{\eta}(t)\end{bmatrix} - \begin{bmatrix}\mathbf{0}\\ \mathbf{C}^T(t)\mathbf{R}^{-1}(t)\mathbf{s}(t)\end{bmatrix}, \qquad T_0 \leqq t \leqq T_f, \tag{4.22}$$

where we have defined

$$\mathbf{W}(t) = \begin{bmatrix}\mathbf{F}(t) & \mathbf{G}(t)\mathbf{Q}\mathbf{G}^T(t)\\ \mathbf{C}^T(t)\mathbf{R}^{-1}(t)\mathbf{C}(t) & -\mathbf{F}^T(t)\end{bmatrix} \tag{4.23}$$

and the boundary conditions

$$\boldsymbol{\eta}(T_f) = \mathbf{0}, \tag{4.18}$$

$$\boldsymbol{\xi}(T_0) = \mathbf{P}_0\,\boldsymbol{\eta}(T_0), \tag{4.19}$$

* This is consistent with the estimator-subtractor realization of the optimal receiver for detecting $\mathbf{s}(t)$. See Ref. 67, pp. 295.

are imposed. Furthermore, let us introduce the following notation. We define the transition matrix of $\mathbf{W}(t)$ to be $\mathbf{\Psi}(t, \tau)$, i.e.,

$$\frac{d}{dt}\mathbf{\Psi}(t, \tau) = \mathbf{W}(t)\mathbf{\Psi}(t, \tau), \tag{4.24a}$$

$$\mathbf{\Psi}(\tau, \tau) = \mathbf{I}. \tag{4.24b}$$

In addition, let us partition this transition matrix into four $n \times n$ submatrices in the form

$$\mathbf{\Psi}(t, \tau) = \begin{bmatrix} \mathbf{\Psi}_{\xi\xi}(t, \tau) & \mathbf{\Psi}_{\xi\eta}(t, \tau) \\ \mathbf{\Psi}_{\eta\xi}(t, \tau) & \mathbf{\Psi}_{\eta\eta}(t, \tau) \end{bmatrix}. \tag{4.25}$$

Method 1

The basic approach of the first method is to use the superposition of a particular and homogeneous solution. First, we generate a convenient particular solution in order to incorporate the forcing term dependence. Then we add a homogeneous solution so as to satisfy the boundary conditions. In order to find a particular solution, let $\xi_p(t)$ and $\boldsymbol{\eta}_p(t)$ be the solution to Equation 4.22 with the initial conditions

$$\xi_p(T_0) = \boldsymbol{\eta}_p(T_0) = \mathbf{0}. \tag{4.26}$$

Since we have specified a complete set of initial conditions we can uniquely solve the equation

$$\frac{d}{dt}\begin{bmatrix} \xi_p(t) \\ \boldsymbol{\eta}_p(t) \end{bmatrix} = \mathbf{W}(t)\begin{bmatrix} \xi_p(t) \\ \boldsymbol{\eta}_p(t) \end{bmatrix} - \begin{bmatrix} \mathbf{0} \\ \mathbf{C}^T(t)\mathbf{R}^{-1}(t)\mathbf{s}(t) \end{bmatrix}, \qquad T_0 \leqq t. \tag{4.27}$$

In order to match the boundary conditions, let us add to this particular solution a homogeneous solution of the form

$$\begin{bmatrix} \xi_h(t) \\ \boldsymbol{\eta}_h(t) \end{bmatrix} = \mathbf{\Psi}(t, T_0)\begin{bmatrix} \mathbf{P}_0 \\ \mathbf{I} \end{bmatrix}\boldsymbol{\eta}(T_0), \tag{4.28}$$

where $\boldsymbol{\eta}_h(T_0)$ is to be chosen. Notice that the sum of these two solutions satisfies the initial boundary condition (Equation 4.13) independent of $\boldsymbol{\eta}_h(T_0)$. Therefore, we want to chose $\boldsymbol{\eta}_h(T_0)$ such that we satisfy the final boundary condition (Equation 4.14). To do this, let us rewrite Equation 4.23 in the form

$$\begin{bmatrix} \xi_h(t) \\ \boldsymbol{\eta}_h(t) \end{bmatrix} = \begin{bmatrix} \mathbf{\Phi}_\xi(t, T_0) \\ \mathbf{\Phi}_\eta(t, T_0) \end{bmatrix}\boldsymbol{\eta}_h(T_0), \tag{4.29}$$

where we define the matrices

$$\boldsymbol{\Phi}_{\xi}(t, T_0) \triangleq \boldsymbol{\Psi}_{\xi\xi}(t, T_0)\mathbf{P}_0 + \boldsymbol{\Psi}_{\xi\eta}(t, T_0), \tag{4.30}$$

$$\boldsymbol{\Phi}_{\eta}(t, T_0) \triangleq \boldsymbol{\Psi}\boldsymbol{\eta}_{\xi}(t, T_0)P(_0 + \boldsymbol{\Psi}_{\eta\eta}(t, T_0). \tag{4.31}$$

Consequently, we have

$$\begin{aligned} \boldsymbol{\xi}(t) &= \boldsymbol{\xi}_p(t) + \boldsymbol{\Phi}_{\eta}(t, T_0)\boldsymbol{\eta}_h(T_0), \\ & \qquad\qquad\qquad\qquad T_0 \leqq t \leqq T_f. \\ \boldsymbol{\eta}(t) &= \boldsymbol{\eta}_p(t) + \boldsymbol{\Phi}_{\eta}(t, T_0)\boldsymbol{\eta}_h(T_0), \end{aligned} \tag{4.32}$$

Applying the final boundary condition requires that

$$\boldsymbol{\eta}_h(T_0) = -\boldsymbol{\Phi}_{\eta}^{-1}(T_f, T_0)\boldsymbol{\eta}_p(T_f). \tag{4.33}$$

Substituting this in Equation 4.28 gives us the final result for this method,

$$\boldsymbol{\xi}(t) = \boldsymbol{\xi}_p(t) - \boldsymbol{\Phi}_{\xi}(t, T_0)\boldsymbol{\Phi}_{\eta}^{-1}(T_f, T_0)\boldsymbol{\eta}_p(T_f), \quad T_0 \leqq t \leqq T_f, \tag{4.34}$$

$$\boldsymbol{\eta}(t) = \boldsymbol{\eta}_p(t) - \boldsymbol{\Phi}_{\eta}(t, T_0)\boldsymbol{\Phi}_{\eta}^{-1}(T_f, T_0)\boldsymbol{\eta}_p(T_f), \quad T_0 \leqq t \leqq T_f. \tag{4.35}$$

(The matrix $\boldsymbol{\Phi}_{\eta}(t, T_0)$ can be shown to be nonsingular for all t.*)

Let us briefly summarize the method. First, we need to determine $\boldsymbol{\Phi}_{\xi}(t, T_0)$ and $\boldsymbol{\Phi}_{\eta}(t, T_0)$ as defined by Equations 4.25, 4.30, and 4.31. (If the state matrices do not depend upon $\mathbf{s}(t)$, this can be done independent of the signal, $\mathbf{s}(t)$.) We then find the particular solutions $\boldsymbol{\xi}_p(t)$ and $\boldsymbol{\eta}_p(t)$ by solving Equation 4.27 with the initial conditions specified by Equation 4.26. Finally, we substitute these functions into Equations 4.32 and 4.33 to find $\boldsymbol{\xi}(t)$ and $\boldsymbol{\eta}(t)$.

Two comments are in order. For a large class of problems the differential equations that one needs to solve using this method have constant coefficients. Consequently, the method is well suited for finding $\boldsymbol{\xi}(t)$ and $\boldsymbol{\eta}(t)$ analytically. We illustrate the use of this method in the next section with two examples.

We also observe that the differential equations we need to solve are unstable, e.g. if the system parameters are constant, $\mathbf{W}$ has eigenvalues with the positive real parts. Consequently, one can (and does) encounter difficulty in numerically solving these equations when the time interval $[T_0, T_f]$ is long. This leads us to the problem of finding an effective numerical procedure for solving one equation.

* See Sec. 3.2 or Ref. 35.

In order to solve this problem we introduce two more methods. The first of these develops some important concepts and results. The second uses concepts to develop the solution method that has the desired numerical properties.

Method 2

The most difficult aspect of solving Equation 4.22 is satisfying the two-point, or mixed, boundary conditions. The essential aspect of the second method is to introduce a third function from which we can determine $\boldsymbol{\xi}(T_f)$. Since $\boldsymbol{\eta}(T_f)$ is always identically zero, this allows us to specify a complete set of boundary conditions at $t = T_f$. With these conditions we then integrate Equation 4.15 backwards over the interval.

From Method 1, let us define

$$\boldsymbol{\Sigma}(t\,|\,t) \triangleq \boldsymbol{\Phi}_{\xi}(t, T_0)\boldsymbol{\Phi}_{\eta}^{-1}(t, T_0). \tag{4.36}$$

We now find a matrix differential equation that $\boldsymbol{\Sigma}(t\,|\,t)$ satisfies. We have

$$\frac{d\boldsymbol{\Phi}_{\xi}(t, T_0)}{dt} = \frac{d\boldsymbol{\Sigma}(t\,|\,t)}{dt}\boldsymbol{\Phi}_{\eta}(t, T_0) + \boldsymbol{\Sigma}(t\,|\,t)\frac{d\boldsymbol{\Phi}_{\eta}(t, T_0)}{dt}, \tag{4.37}$$

Substituting from Equations 4.30, 4.31, and 4.25, we find

$$\begin{aligned}\mathbf{F}(t)\boldsymbol{\Phi}_{\xi}(t, T_0) &+ \mathbf{G}(t)\mathbf{Q}\mathbf{G}^T(t)\boldsymbol{\Phi}_{\eta}(t, T_0)\\ &= \frac{d\boldsymbol{\Sigma}(t\,|\,t)}{dt}\boldsymbol{\Phi}_{\eta}(t, T_0)\\ &\quad + \boldsymbol{\Sigma}(t\,|\,t)(\mathbf{C}^T(t)\mathbf{R}^{-1}(t)\mathbf{C}(t)\boldsymbol{\Phi}_{\xi}(t, T_0) - \mathbf{F}^T(t)\boldsymbol{\Phi}_{\eta}(t, T_0)).\end{aligned} \tag{4.38}$$

Multiplying by $\boldsymbol{\Phi}_{\eta}^{-1}(t, T_0)$ and using Equation 4.26 yields

$$\begin{aligned}\frac{d\boldsymbol{\Sigma}(t\,|\,t)}{dt} = \mathbf{F}(t)\boldsymbol{\Sigma}(t\,|\,t) + \boldsymbol{\Sigma}(t\,|\,t)\mathbf{F}^T(t) - \boldsymbol{\Sigma}(t\,|\,t)\mathbf{C}^T(t)\mathbf{R}^{-1}(t)\mathbf{C}(t)\boldsymbol{\Sigma}(t\,|\,t)\\ + \mathbf{G}(t)\mathbf{Q}\mathbf{G}^T(t).\end{aligned} \tag{4.39}$$

The initial condition follows from Equations 4.30 and 4.31

$$\boldsymbol{\Sigma}(T_0\,|\,T_0) = \mathbf{P}_0. \tag{4.39}$$

Consequently, we have the expected result that $\boldsymbol{\Sigma}(t\,|\,t)$ is identical with the realizable filter covariance matrix since it satisfies the same matrix Ricatti differential equation and has the same initial condition.*

* See the discussion in Ref. 35 regarding the solution of the variance equation.

Let us define a function $\xi_r(t)$

$$\xi_r(t) = \xi_p(t) - \Phi_\xi(t, T_0)\Phi_\eta^{-1}(t, T_0)\eta_p(t) = \xi_p(t) - \Sigma(t \mid t)\eta_p(t). \quad (4.41)$$

We note that

$$\xi(T_f) = \xi_r(T_f). \quad (4.42)$$

Now we find a differential equation for $\xi_r(t)$. Differentiating Equation 4.41 and using Equation 4.39, yields

$$\begin{aligned}\frac{d\xi_r(t)}{dt} = \mathbf{F}(t)\xi_p(t) + \mathbf{G}(t)\mathbf{Q}\mathbf{G}^T(t)\eta_p(t) \\ -(\mathbf{F}(t)\Sigma(t \mid t) + \Sigma(t \mid t)\mathbf{F}^T(t) + \mathbf{G}(t)\mathbf{Q}\mathbf{G}^T(t) \\ -\Sigma(t \mid t)\mathbf{C}^T(t)\mathbf{R}^{-1}(t)\mathbf{C}(t)\Sigma(t \mid t))\eta_p(t) \\ -\Sigma(t \mid t)((\mathbf{C}^T(t)\mathbf{R}^{-1}(t)\mathbf{C}(t)\xi_p(t) - \mathbf{F}^T(t)\eta_p(t) \\ -\mathbf{C}^T(t)\mathbf{R}^{-1}(t)\mathbf{s}(t))\end{aligned} \quad (4.43)$$

After canceling and combining terms by using Equation 4.41 we have

$$\frac{d\xi_r(t)}{dt} = \mathbf{F}(t)\xi_r(t) + \Sigma(t \mid t)\mathbf{C}^T(t)\mathbf{R}^{-1}(t)(\mathbf{s}(t) - \mathbf{C}(t)\xi_r(t)). \quad (4.44)$$

The initial condition follows from Equations 4.41, 4.36, and 4.27

$$\xi_r(T_0) = \mathbf{0}. \quad (4.45)$$

From Equation 4.43 we have the expected result that $\xi_r(t)$ satisfies a differential equation of the same form as realizable filter estimator equation. We should note that in this particular application of solving the inhomogeneous integral equation, the input $\mathbf{s}(t)$ is deterministic rather than a random process, e.g., some received signal $\mathbf{r}(t)$.

Equations 4.39 and 4.43 are the key to the second method. We simply integrate them forward in time to $t = T_f$, apply Equation 4.42 to find $\xi(T_f)$, and then integrate Equation 4.20 backwards in time using the complete set of boundary conditions at $t = T_f$. Expressed in terms of an integral operation we have

$$\begin{bmatrix}\xi(t) \\ \eta(t)\end{bmatrix} = \Psi(t, T_f)\begin{bmatrix}\xi_r(T_f) \\ \mathbf{0}\end{bmatrix} + \int_t^{T_f} \Psi(t, t')\begin{bmatrix}\mathbf{0} \\ \mathbf{C}^T(t')\mathbf{R}^{-1}(t')\mathbf{s}(t')\end{bmatrix} dt',$$

$$T_0 \leqq t \leqq T_f. \quad (4.46)$$

Let us examine this method for a moment. The basic approach was to convert a two-point boundary problem into an initial, or final, value problem. Since the $\xi_r(t)$ function that we developed for this conversion is the output of a realizable filter structure, it has many desirable properties. In particular, much is known regarding effective procedures for calculating $\xi_r(t)$ numerically.

However, we still have difficulty integrating Equation 4.46 backwards in time since Equation 4.21 is also unstable when integrated backwards. For example, with a constant-parameter system $\mathbf{W}$ has eigenvalues in the left half-plane. These produce growing exponentials as we integrate Equation 4.17 backwards from the end point. With our third method we eliminate this undesirable feature.

Method 3

In this method we shall derive a result that allows us to uncouple the equations for $\xi(t)$ and $\boldsymbol{\eta}(t)$. After we do this we shall observe that the resulting differential equations for $\xi(t)$ and $\boldsymbol{\eta}(t)$ have some desirable features from a computational viewpoint. First we need to derive one key result.

Let us consider the difference between $\xi(t)$ and $\xi_r(t)$. Substituting Equations 4.34 and 4.41

$$\begin{aligned}\xi(t) - \xi_r(t) &= \boldsymbol{\Phi}_{\xi}(t, T_0)\{-\boldsymbol{\Phi}_{\boldsymbol{\eta}}^{-1}(T_f, T_0)\boldsymbol{\eta}_p(T_f) + \boldsymbol{\Phi}_{\boldsymbol{\eta}}^{-1}(t, T_0)\boldsymbol{\eta}_p(t)\}\\ &= \boldsymbol{\Phi}_{\xi}(t, T_0)\boldsymbol{\Phi}_{\boldsymbol{\eta}}^{-1}(t, T_0)\{\boldsymbol{\eta}_p(t) - \boldsymbol{\Phi}_{\boldsymbol{\eta}}(t, T_0)\\ &\qquad \boldsymbol{\Phi}_{\boldsymbol{\eta}}^{-1}(T_f, T_0)\boldsymbol{\eta}_p(T_f)\}\\ &= \boldsymbol{\Sigma}(t \mid t)\boldsymbol{\eta}(t). \end{aligned} \tag{4.47}$$

Consequently, we have the result

$$\xi(t) - \xi_r(t) = \boldsymbol{\Sigma}(t \mid t)\boldsymbol{\eta}(t) \tag{4.48}$$

or

$$\boldsymbol{\eta}(t) = \boldsymbol{\Sigma}^{-1}(t \mid t)(\xi(t) - \xi_r(t)). \tag{4.49}$$

If we substitute this into Equation 4.14 we can obtain separately the differential equations for $\xi(t)$ and $\boldsymbol{\eta}(t)$. We find, substituting for $\boldsymbol{\eta}(t)$,

$$\begin{aligned}\frac{d\xi(t)}{dt} &= \mathbf{F}(t)\xi(t) + \mathbf{G}(t)\mathbf{Q}\mathbf{G}^T(t)\boldsymbol{\Sigma}^{-1}(t \mid t)(\xi(t) - \xi_r(t))\\ &= (\mathbf{F}(t) + \mathbf{G}(t)\mathbf{Q}\mathbf{G}^T(t)\boldsymbol{\Sigma}^{-1}(t \mid t))\xi(t)\\ &\qquad - \mathbf{G}(t)\mathbf{Q}\mathbf{G}^T(t)\boldsymbol{\Sigma}^{-1}(t \mid t)\xi_r(t), \qquad T_0 \leqq t \leqq T_f. \end{aligned} \tag{4.50}$$

Similarly, substituting for $\xi(t)$ yields

$$\frac{d\eta(t)}{dt} = \mathbf{C}^T(t)\mathbf{R}^{-1}(t)\mathbf{C}(t)(\Sigma(t\,|\,t)\eta(t) + \xi_r(t))$$

$$-\mathbf{F}^T(t)\eta(t) - \mathbf{C}^T(t)\mathbf{R}^{-1}(t)\mathbf{s}(t)$$

$$= -(\mathbf{F}(t) - \Sigma(t\,|\,t)\mathbf{C}^T(t)\mathbf{R}^{-1}(t)\mathbf{C}(t))^T\eta(t)$$

$$-\mathbf{C}^T(t)\mathbf{R}^{-1}(t)(\mathbf{s}(t) - \mathbf{C}(t)\xi_r(t)), \qquad T_0 \leqq t \leqq T_f. \tag{4.51}$$

We now note that by finding $\xi_r(t)$ we can solve either Equation 4.50 or Equation 4.51 for $\xi(t)$ or $\eta(t)$, respectively. The initial (or final) condition for Equation 4.50 is given by Equation 4.42, while for Equation 4.51 it is given by Equation 4.19. Either function can be obtained from the other by using Equation 4.48

Now let us examine the stability aspects of these equations when integrated backwards. Our discussion is essentially qualitative. First, we need to examine the coefficient matrices of Equation 4.45 and Equation 4.46. In Equation 4.46, this matrix is

$$-(\mathbf{F}(t) - \Sigma(t\,|\,t)\mathbf{C}^T(t)\mathbf{R}^{-1}(t)\mathbf{C}(t))^T,$$

which is the negative transpose of the coefficient matrix of the realizable filter. Consequently, if it is stable, so is Equation 4.51 when integrated backwards.

For constant-parameter systems, we can see heuristically that Equation 4.51 is also stable when integrated backwards over large time intervals. If the interval is "long,"

$$d\Sigma(t\,|\,t)/dt \cong 0 \tag{4.52}$$

over most of the interval. If we assume equality, i.e., $\Sigma(t\,|\,t) = \Sigma_\infty$, we have

$$\mathbf{F} + \mathbf{GQG}^T\Sigma_\infty^{-1} = -\Sigma_\infty[-(\mathbf{F} - \Sigma_\infty \mathbf{C}^T\mathbf{R}^{-1}\mathbf{C})^T]\Sigma_\infty^{-1}; \tag{4.53}$$

or, $\mathbf{F} + \mathbf{GQG}\Sigma_\infty^{-1}$ is similar to $-(\mathbf{F} - \Sigma_\infty \mathbf{C}^T\mathbf{R}^{-1}\mathbf{C})^T$. This implies that both matrices have the same eigenvalues; therefore, Equation 4.50 is similarly stable when integrated backwards over the interval. Consequently, one can solve numerically either Equation 4.50 or Equation 4.51 and obtain stable solutions. However, if $\Sigma(t\,|\,t)$ is also available we should point out that Equation 4.50 requires its inversion whereas Equation 4.51 does not.

Summary of Methods

In this section we have developed in considerable detail methods that exist for solving the differential equations we derived for the inhomogeneous Fredholm integral equation. Let us now summarize these methods by discussing their applicability.

If we have a constant-parameter system and want to find an analytical solution, method 1 is probably most useful, since the differential equations have constant coefficients for a large class of problems. However, if we want to obtain a numerical solution, especially over a long time, method 3 is probably the best, since methods 1 and 2 can create some difficulties when integrated numerically. We really never use method 2. The essential reason for introducing it is that it was an intermediate result in our derivation of method 3.

We now apply the results of this section to analyze some examples of solving inhomogeneous Fredholm equations with our method developed in Section 4.2.

4.4 Examples of Solutions to the Inhomogeneous Fredholm Equation

In this section we consider three examples to illustrate the results of the previous two sections. Again, we work two examples analytically while we use numerical procedures for the third. In those examples that we work analytically, we use method 1 as discussed in the last section. The computer program used in the numerical example integrated the second differential equation of method 3.

We present the examples in the context of detection in colored noise. After determining the solution $\mathbf{g}(t)$, we compute the d^2 (and d_g^2) performance measures discussed at the end of the previous section. Finally, in all three examples we set $T_0 = 0$ and $T_f = T$ and assume that the forcing function, or transmitted signal, $s(t)$ is a pulsed sine wave with unit energy over this interval, i.e.,

$$s(t) = \sqrt{\frac{2}{T}} \sin\left(\frac{n\pi t}{T}\right), \qquad 0 \leqq t \leqq T. \tag{5.54}$$

We also assume that $\mathbf{R}(t)$, or the white noise level, is a scalar constant

$$\mathbf{R}(t) = \sigma > 0. \tag{4.55}$$

Example 4.1 g(t) for a Wiener Process

Let us consider the problem of finding $g(t)$ when the kernel is the covariance of a Wiener process. Equations 2.23 and 2.24 describe a system for generating this process. First, we substitute the parameters of these equations into Equation 4.17 in order to find the equation that we need to solve. We obtain

$$\frac{d}{dt}\begin{bmatrix}\xi(t)\\ \eta(t)\end{bmatrix} = \begin{bmatrix}0 & 1\\ \frac{\mu^2}{\sigma} & 0\end{bmatrix}\begin{bmatrix}\xi(t)\\ \eta(t)\end{bmatrix} - \begin{bmatrix}0\\ \frac{\mu}{\sigma}\sqrt{\frac{2}{T}}\sin\left(\frac{n\pi t}{T}\right)\end{bmatrix},$$

$$0 \leqq t \leqq T. \tag{4.56}$$

From Equation 4.18 and Equation 4.19 the boundary conditions are

$$\boldsymbol{\xi}(0) = 0, \tag{4.57}$$

$$\boldsymbol{\eta}(T) = 0. \tag{4.58}$$

Referring to the previous section, method 1, we find particular solutions $\boldsymbol{\xi}_p(t)$ and $\boldsymbol{\eta}(t)$. These are the solutions to Equation 4.17 with $\boldsymbol{\xi}_p(0) = \boldsymbol{\eta}_p(0) = 0$. Doing this we obtain

$$\begin{bmatrix}\xi_p(t)\\ n_p(t)\end{bmatrix} = \frac{\mu}{\sigma\gamma^2}\sqrt{\frac{2}{T}}\begin{bmatrix}\sin\left(\frac{n\pi t}{T}\right)\\ \left(\frac{n\pi}{T}\right)\cos\left(\frac{n\pi t}{T}\right)\end{bmatrix}, \tag{4.59}$$

where, for this problem, we define γ^2 to be

$$\gamma^2 = \left[\left(\frac{n\pi}{T}\right)^2 + \frac{\mu^2}{\sigma}\right]. \tag{4.60}$$

Next, we need to find the transition matrix associated with Equation 4.17. After some straightforward calculation we find

$$\boldsymbol{\Psi}(t, 0) = \begin{bmatrix}\cosh\left[\left(\frac{\mu^2}{\sigma}\right)^{1/2} t\right] & \left[\frac{\mu^2}{\sigma}\right]^{-1/2}\sinh\left[\left(\frac{\mu^2}{\sigma}\right]^{1/2} t\right)\\ \left[\frac{\mu^2}{\sigma}\right]^{1/2}\sinh\left(\left[\frac{\mu^2}{\sigma}\right]^{1/2} t\right) & \cosh\left(\left[\frac{\mu^2}{\sigma}\right]^{1/2} t\right)\end{bmatrix}. \tag{4.61}$$

We need to add a homogeneous equation of the form of Equation 4.28

$$\begin{bmatrix}\xi_h(t)\\ \eta_h(t)\end{bmatrix} = \begin{bmatrix}\left[\frac{\mu^2}{\sigma}\right]^{-1/2}\sinh\left(\left[\frac{\mu^2}{\sigma}\right]^{1/2} t\right)\\ \cosh\left(\left[\frac{\mu^2}{\sigma}\right]^{1/2} t\right)\end{bmatrix}\eta_h(0). \tag{4.62}$$

From Equations 4.62 and 4.59 we find that we must choose $\boldsymbol{\eta}_h(0)$ to be Equation 4.33

$$\eta_h(0) = -\frac{\mu}{\sigma\gamma^2}\sqrt{\frac{2}{T}}\left(\frac{n\pi}{T}\right)(-1)^n \bigg/ \cosh\left(\left[\frac{\mu^2}{\sigma}\right]^{1/2} T\right). \tag{4.63}$$

Consequently, we have

$$\xi(t) = \frac{\mu}{\sigma\gamma^2}\sqrt{\frac{2}{T}}\left\{\sin\left(\frac{n\pi t}{T}\right) - \left(\frac{n\pi}{T}\right)(-1)^n\left[\frac{\mu^2}{\sigma}\right]^{-1/2} \times \left[\frac{\sinh([\mu^2/\sigma]^{1/2}t)}{\cosh([\mu^2/\sigma]^{1/2}T)}\right]\right\}, \quad 0 \leqq t \leqq T. \tag{4.64}$$

The solution $\mathbf{g}(t)$ is found by substituting Equation 4.64 into Equation 4.13

$$g(t) = \frac{1}{\sigma}\sqrt{\frac{2}{T}}\left(\frac{n\pi}{T\gamma}\right)^2\left\{\sin\left(\frac{n\pi t}{T}\right) + \left(\left[\frac{\mu^2}{\sigma}\right]^{1/2}\left(\frac{n\pi}{T}\right)^{-1}\right)(-1)^n \times \left[\frac{\sinh([\mu^2/\sigma]^{1/2}t)}{\cosh([\mu^2/\sigma]^{1/2}T)}\right]\right\}, \quad 0 \leqq t \leqq T. \tag{4.65}$$

After some straightforward but tedious manipulation we can also calculate d_g^2 as given by Equation 4.21c

$$d_g^2 = \frac{\mu^2}{\sigma^2\gamma^2}\left\{1 + \frac{2}{\gamma^2}\left(\frac{n\pi}{T}\right)^2 \frac{\tanh([\mu^2/\sigma]^{1/2}T)}{[\mu^2/\sigma]^{1/2}T}\right\}. \tag{4.66}$$

Example 4.2 g(t) for a First-Order Stationary Spectrum

Let us consider the kernel to be the covariance function for a one-pole stationary process as described by Equations 2.25 and 2.26. From these equations and Equation 4.17, the differential equations that we need to solve are

$$\frac{d}{dt}\begin{bmatrix}\xi(t)\\ \eta(t)\end{bmatrix} = \begin{bmatrix}-k & 2kP\\ 1/\sigma & k\end{bmatrix}\begin{bmatrix}\xi(t)\\ \eta(t)\end{bmatrix} - \begin{bmatrix}0\\ \frac{1}{\sigma}\sqrt{\frac{2}{T}}\sin\left(\frac{n\pi t}{T}\right)\end{bmatrix}, \quad 0 \leqq t \leqq T, \tag{4.67}$$

subject to the boundary conditions specified by Equations 4.18 and 4.19. We now use the first solution method that we discussed. After some manipulation, we find

$$\xi_p(t) = \frac{P}{\gamma'^2 \Lambda \sigma}\sqrt{\frac{2}{T}}\left[2k\Lambda \sin\left(\frac{n\pi t}{T}\right) - \left(\frac{n\pi}{T}\right)(e^{k\Lambda t} - e^{-k\Lambda t})\right], \qquad 0 < t, \tag{4.68}$$

$$\eta_p(t) = \frac{1}{\gamma'^2 \sigma}\sqrt{\frac{2}{T}}\left[k \sin\left(\frac{n\pi t}{T}\right) + \left(\frac{n\pi}{T}\right)\cos\left(\frac{n\pi t}{T}\right) - \left(\frac{n\pi}{T}\right)\left[\frac{1}{2}\left(1 + \frac{1}{\Lambda}\right)e^{k\Lambda t}\right] - \left(\frac{n\pi}{T}\right)\left[\frac{1}{2}\left(1 - \frac{1}{\Lambda}\right)e^{-k\Lambda t}\right]\right], \qquad 0 < t, \tag{4.69}$$

where

$$\Lambda = \sqrt{1 + \frac{2P}{k\sigma}}, \tag{4.70}$$

$$\gamma'^2 = \left[\left(\frac{n\pi}{T}\right)^2 + (k\Lambda)^2\right]. \tag{4.71}$$

To compute the homogenous solution we need to find the transition matrix $\mathbf{\Psi}(t, 0)$. We do this using the Laplace transform approach and obtain

$$\mathbf{\Psi}(t, 0) = \begin{bmatrix} \frac{1}{2}\left(1 - \frac{1}{\Lambda}\right)e^{k\Lambda t} + \frac{1}{2}\left(1 + \frac{1}{\Lambda}\right)e^{-k\Lambda t} & \frac{P}{\Lambda}(e^{k\Lambda t} - e^{-k\Lambda t}) \\ \frac{1}{2k\Lambda\sigma}(e^{k\Lambda t} - e^{-k\Lambda t}) & \frac{1}{2}\left(1 + \frac{1}{\Lambda}\right)e^{k\Lambda t} + \frac{1}{2}\left(1 - \frac{1}{\Lambda}\right)e^{-k\Lambda t} \end{bmatrix}. \tag{4.72}$$

Next we find $\mathbf{\Phi}_{\boldsymbol{\xi}}(t, 0)$ and $\mathbf{\Phi}_{\boldsymbol{\eta}}(t, 0)$ as specified by Equations 4.34 and 4.35. When we add these homogeneous solutions to the above expressions for $\boldsymbol{\xi}_p(t)$ and $\boldsymbol{\eta}_p(t)$, we find

$$\xi(t) = \xi_p(t) + \frac{P}{2}\left[\left(1 + \frac{1}{\Lambda}\right)e^{k\Lambda t} + \left(1 - \frac{1}{\Lambda}\right)e^{-k\Lambda t}\right]\eta_h(0), \qquad 0 \leqq t \leqq T, \tag{4.73}$$

$$\eta(t) = \eta_p(t) + \frac{1}{4\Lambda}[(\Lambda + 1)^2 e^{k\Lambda t} - (\Lambda - 1)^2 e^{-k\Lambda t}]\eta_h(0), \qquad 0 \leqq t \leqq T, \tag{4.74}$$

where

$$\boldsymbol{\eta}_h(0) = -\boldsymbol{\Phi}_{\boldsymbol{\eta}}^{-1}(T, 0)\boldsymbol{\eta}_p(T) = -\left[\frac{(\Lambda+1)^2 e^{k\Lambda T} - (\Lambda-1)^2 e^{-k\Lambda T}}{4\Lambda}\right]^{-1}$$

$$\times \frac{1}{\gamma'^2\sigma}\sqrt{\frac{2}{T}}\left(\frac{n\pi}{T}\right)\left[(-1)^n - \frac{1}{2}\left(1+\frac{1}{\Lambda}\right)e^{k\Lambda T} - \frac{1}{2}\left(1-\frac{1}{\Lambda}\right)e^{-k\Lambda T}\right]. \tag{4.75}$$

When we make all the substitutions required, we finally obtain

$$g(t) = \sqrt{\frac{2}{T}}\frac{1}{\gamma'^2\sigma}\left\{\left[\left(\frac{n\pi}{T}\right)^2 + k^2\right]\sin\left(\frac{n\pi t}{T}\right)\right.$$

$$+\frac{2Pn\pi}{T}[(\Lambda+1)^2 e^{k\Lambda T} - (\Lambda-1)^2 e^{-k\Lambda T}]^{-1}$$

$$\times ((-1)^n[(\Lambda+1)e^{k\Lambda t} + (\Lambda-1)e^{-k\Lambda t}]$$

$$\left. - [(\Lambda+1)e^{k\Lambda(T-t)} + (\Lambda-1)e^{-k\Lambda(T-t)}])\right\}, \qquad 0 \leqq t \leqq T. \tag{4.76}$$

One can continue and evaluate the performance by computing d^2 and d_g^2 according to Equation 4.23; however, the result is rather complex and is not informative. Instead of presenting an analytic formula, we plot d^2 and d_g^2 against n for a particular choice of parameters. The results are presented in Figure 4.2 when $k=1$, $\sigma=1$, $P=1$, and $T=2$. For the case $n=1$, we see that the presence of the colored noise degrades our performance approximately 50 per cent from that of the white noise. For $n=8$, however, our performance is within 2 per cent of the performance for the white-noise-only performance. This is what we would intuitively expect. As n increases, the signal bandwidth increases and moves outside the region where the colored component of the noise has significant power. For $n=8$ we are very close to a matched filter receiver.

It should be apparent from the complexity of these two simple examples that an analytic solution for $g(t)$ and the associated performance is indeed a very difficult task when the kernel is the covariance of a process generated by higher-order system. Consequently, it is desirable to have an efficient numerical method.

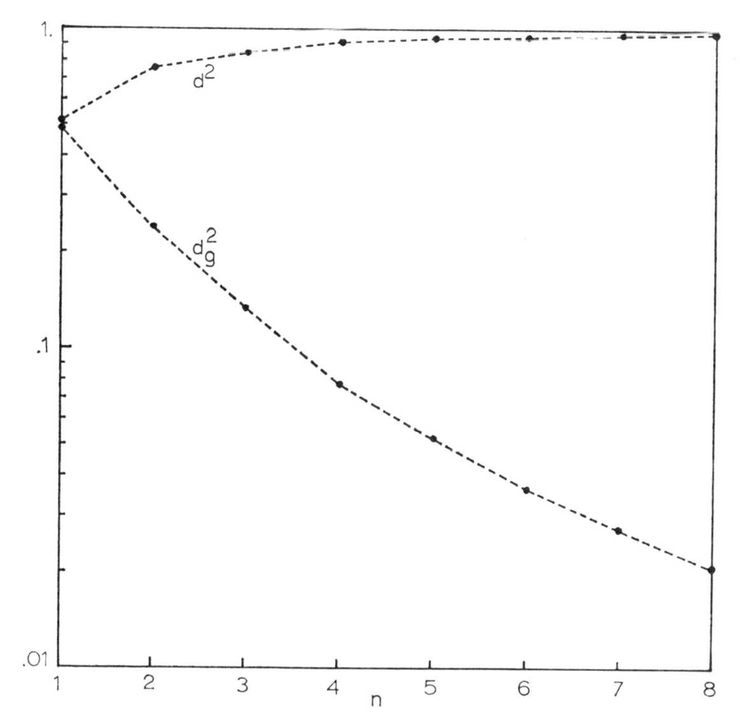

Figure 4.2 Performance d^2 and degradation $d_g^2 = d_w^2 - d^2$ for first-order noise with $s(t) = \sin(n\pi t)$, $0 \leq t \leq T$.

Example 4.3 $g(t)$ **for a Second-Order Stationary Process**

For the third example, we consider a numerical approach to the analysis of a second-order system. We chose $K_y(t, \tau)$ to be the covariance generated by the state description of Equations 2.28 and 2.29. We consider three levels of the white noise, $\sigma = 1/10$, 1, 10. The interval $[0, T]$ is specified by setting $T = 2$.

We find $g(t)$ using method 3 in the previous section. We integrate Equations 4.39 and 4.44. forward in time using the initial conditions of Equations 4.40 and 4.45, respectively. Note that Equations 4.39 are not coupled to the choice of the signal if the coefficient matrices in the state description are not; consequently, for this case it can be solved independent of the particular value of n. We integrate Equation 4.51 backwards to find $\boldsymbol{\eta}(t)$, and then use Equations 4.48 and 4.13 to determine $\mathbf{g}(t)$.

The results for $n = 1$ and $n = 2$ for the various values of s are presented in Figures 4.3 and 4.4. It is useful to compare the results in the context

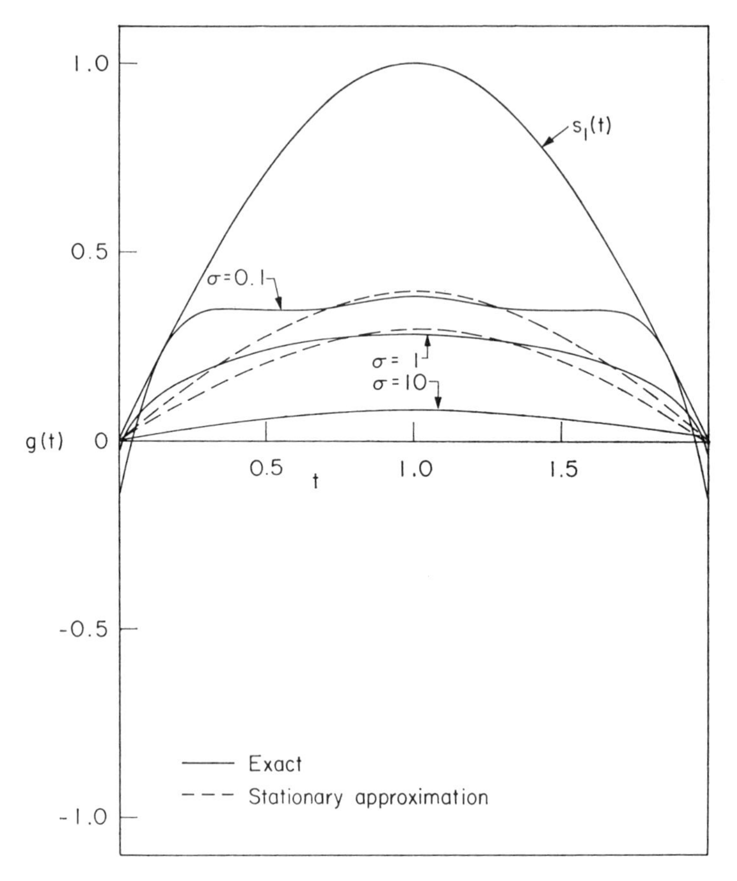

Figure 4.3 The solution to the inhomogeneous equation $g(t)$ for $s(t) = \sin(\pi t/2)$.

of the relative bandwidths of the signal and the colored noise process. We use an rms definition of bandwidth for the signal

$$B_n = \frac{1}{2\pi}\left[\frac{\int_{-\infty}^{\infty} \omega^2 |S_{s_n}(\omega)|^2 \, d\omega}{\int_{-\infty}^{\infty} |S_{s_n}(\omega)|^2 \, d\omega}\right]^{1/2} H_z; \tag{4.77}$$

consequently, we have for our choice of signals

$$B_n = n/4H_z.$$

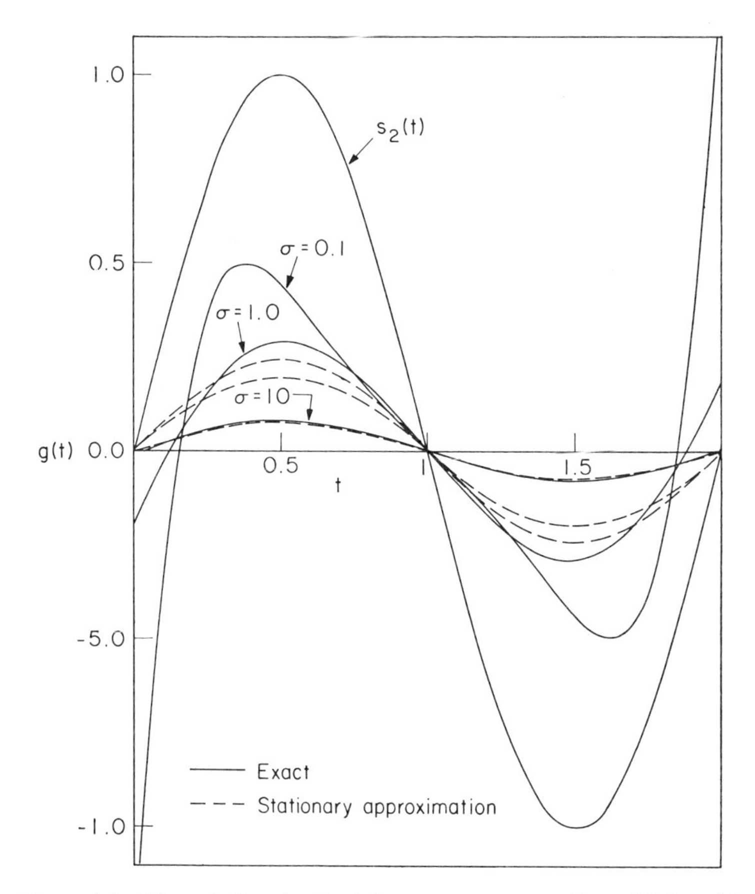

Figure 4.4 The solution to the inhomogeneous equation $g(t)$ for $s(t) = \sin(\pi t)$.

For this choice of parameters the spectrum of $s_n(t)$ and the power density spectrum $S_y(\omega)$ overlap significantly. We have also indicated the results that a long-time stationary process analysis yield. We can see that, for this example, the end effects are definitely significant when the colored component of the noise is dominant ($\sigma = 1/10$). This is consistent with the appearance of impulses when $\sigma = 0$.* With $\sigma = 10$, the results do not differ too much from a matched filter result.

We have summarized the performance results for this example in Figure 4.5 by plotting the normalized parameters d^2/d_w^2 and d_g^2/d_w^2 versus

* Ref. 22, pp. 240–242.

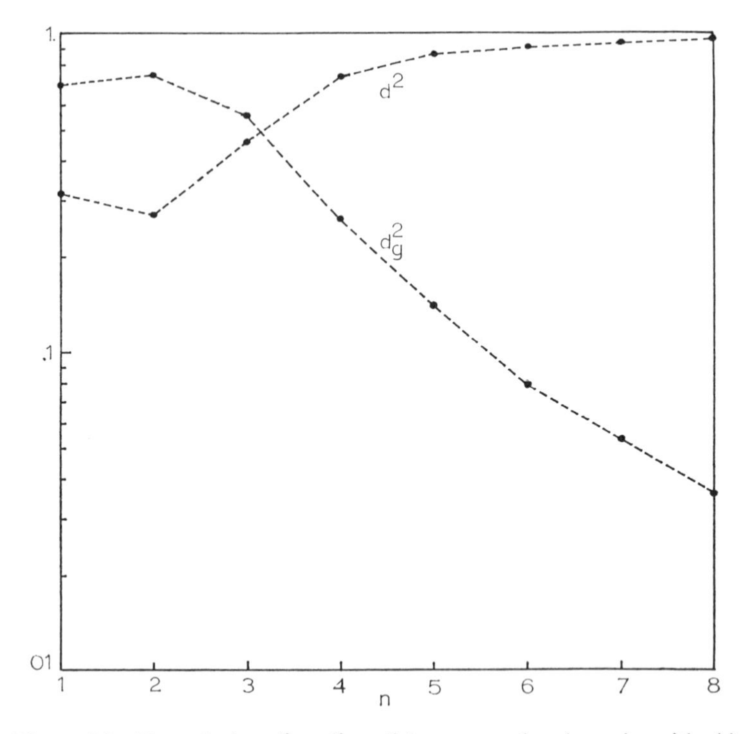

Figure 4.5 Degradation $d_g^2 = d_w^2 - d^2$ for a second-order noise with $s(t) = \sin(n\pi t)$, $0 \leq t \leq T$.

n. We can see that the colored noise has the most significant effect when $n = 2$. For specific parameters choices this corresponds to a bandwidth of 1/2, which is close to the frequency where the colored noise has its maximum value, and is where the transient effects at the end points are most significant. As we move past the peak, the performance improves monotonically with n.

4.5 Discussion of Results for Inhomogeneous Fredholm Integral Equations

In this chapter we have formulated a state variable approach for the solution of inhomogeneous Fredholm integral equations. Let us briefly compare our approach to some of the existing ones.

The approach of reducing an integral equation to a differential equation certainly is not new.[55] In one form or another it is undoubtedly the most common procedure used. In comparison to other differential equation methods, our approach has several advantages (many of which are shared with our solution method for the homogeneous equation).

1. We can solve Equation 4.4 when $\mathbf{s}(t)$ is a vector function.
2. The differential equations that must be solved follow directly once the state matrices that describe the generation of the kernel are chosen.
3. We do not have to substitute any functions back into the original integral equation in order to find a set of linear equations that must be solved.
4. We can study a wide class of time-varying systems.
5. The technique is well suited for numerical solutions finding.

There are two major disadvantages.

1. The class of kernels that may be considered is limited. However, the technique is applicable to a large and important class of kernels that appear in communications.
2. We cannot handle integral equations of the first kind, e.g., when the white component of the noise is identically zero. For these equations singularity functions appear at the interval end points. We excluded these in our derivation. We should note that we have observed the limiting behavior of our solution approaching these singularity functions when the white noise is small.

A second method that is in contrast to the differential equation approaches is to find the inverse kernel of the integral equation. The inverse kernel $\mathbf{Q}(\tau, u)$ is defined so that it satisfies a second integral equation

$$\int_{T_0}^{T_f} \{\mathbf{R}(t)\,\delta(t-\tau) + \mathbf{K}_y(t, \tau)\}\mathbf{Q}(\tau, u)\,d\tau = \mathbf{I}\delta(t-u), \qquad T_0 < t, u < T_f. \tag{4.78}$$

In terms of the inverse kernel the solution $\mathbf{g}(t)$ is found to be

$$\mathbf{g}(t) = \int_{T_0}^{T_f} \mathbf{Q}(t, u)\mathbf{s}(u)\,du, \qquad T_0 < t < T_f. \tag{4.79}$$

One common numerical method that uses this approach is to approximate the integral operations 4.78 and 4.79 by matrices

$$[\mathbf{K}][\mathbf{Q}] = \mathbf{I}, \tag{4.80}$$

$$\mathbf{g} = [\mathbf{Q}]\mathbf{s}. \tag{4.81}$$

Let us briefly compare the computation required using this approach to that of our approach. If we assume that we sample the interval at

NI points, Equations 4.79 and 4.80 are *NI* dimensional. One can show that the number of computations required to find $[\mathbf{Q}]$ as given by 4.79 goes as (*NI*) cubed. If we assume that we find $[\mathbf{Q}]$ by specifying the coefficients of our differential equations, the computations we require increases only linearly with *NI*. The computation required to implement Equation 4.79 is proportional to *NI* squared, whereas the computations required to solve our equations again is linearly proportional to *NI*. The conclusion is that, for large *NI*, which is required for high accuracy, the differential equation approach is superior.

Before we leave the topic of the inverse kernel, we point out an important concept that we use in a later chapter. We can consider that the inhomogeneous integral equation specifies a linear operation. In an explicit integral representation, this linear operation is given by Equation 4.78. It is completely equivalent, however, to specify $\mathbf{g}(t)$ implicitly as the solution to our differential equations.

In the two following chapters we shall apply the results of this chapter. In Chapter 5 we apply them to the problem of designing optimal signals for detection in additive colored noise channels. Our basic approach is to regard the differential equations we have developed as a dynamic system with initial and final boundary conditions. When the problem is expressed in this form we can apply the maximal principle of Pontryagin for the optimization. In Chapter 6 we present a new approach to solving Wiener-Hopf equations by using the results of this chapter. We then proceed to develop a unified theory of linear smoothing and filtering with delay.

5 Optimal Signal Design for Colored Noise Channels via State Variables

Many channels of interest in detection problems have the property of producing a colored, or nonwhite, noise at the receiver terminals in addition to the signal and the usual white noise. The presence of this nonwhite noise components leads to the important characteristic that the transmitted signal shape affects the performance of the system. For a given amount of energy, or power, certain signal waveforms result in lower error probabilities than do others. Consequently, by choosing the signal waveform in some optimal manner, we may maximize the system performance.

In general, the signal optimization must be performed within certain constraints. One always has some type of amplitude constraints imposed by the physical aspects of the transmitter. These quite often take the form of peak amplitude and/or energy constraints. It is also realistic to impose bandwidth constraints. These may also arise from the physical aspects of the transmitter or from adjacent channel considerations. The precise form of the constraint may take several forms depending upon the definition of bandwidth used and its relevance to a given problem. The net effect of these bandwidth constraints is that we are often prevented from putting the signal energy in frequency bands disjoint from the noise, e.g., on the "tail" of the colored noise spectrum.

5.1 Problem Formulation

In Chapter 4 we introduced the problem of detecting signals with a known waveform in the presence of an additive colored noise. The system model is illustrated in Figure 4.1. It leads to the following simple binary detection problem*†:

$$\begin{aligned} &\text{on } H_1: \quad r(t) = s(t) + y(t, s(\cdot)) + w(t), \quad T_0 \leqq t \leqq T_f, \\ &\text{on } H_0: \quad r(t) = y(t, s(\cdot)) + w(t), \qquad\qquad T_0 \leqq t \leqq T_f, \end{aligned} \tag{5.1}$$

where $s(t)$ is the signal to be optimized, $y(t:s(\cdot))$ is a colored noise process which may depend upon the transmitted signal, and $w(t)$ is an independent additive white noise process of spectral height $N_0/2$.

In the simplest case, the colored noise is simply additive and independent of the transmitted signal. One also often encounters situations where the transmitted signal strongly influences the properties of the noise, for example, in detecting a point target in a reverberating environment. Whereas in the first case the optimization problem is linear and is relatively easy to perform, in the second the problem becomes nonlinear, and iterative methods need to be introduced.

As discussed in Chapter 4, the optimal receiver is a correlator. The correlating waveform $g(t)$ and the performance measure d^2 is specified by solving a Fredholm integral equation. The correlating signal $g(t)$ for this problem is the solution of the equation

$$\int_{T_0}^{T_f} K_y(t, \tau) g(\tau)\, d\tau + \frac{N_0}{2} g(t) = s(t), \qquad T_0 \leqq t \leqq T_f, \tag{5.2}$$

where $s(t)$ is the transmitted signal, $g(t)$ is the optimal correlating signal, $N_0/2$ is the white-noise level (identified as $\mathbf{R}(t)$ in the general vector case), and

$$K_y(t, \tau) = E[y(t:s(\cdot))y(\tau:s(\cdot))]$$

is the covariance function of the colored (finite power) component of the additive noise. The performance measure d^2 is given by

$$d^2 = \int_{T_0}^{T_f} s(t) g(t)\, dt. \tag{5.3}$$

* We also allowed $s(t)$ to be multiplied by a random coefficient b. This led to a similar formulation the only change being the way ROC's are computed using the parameter d^2. Cf. Eqs. (4.7)–(4.9).

† In this chapter we confine our attention to scalar channels.

Since practical constraints are usually imposed, the choice of the signal $s(t)$ is not completely free. We impose two constraints upon the signal. The first is an energy constraint, or

$$\int_{T_0}^{T_f} s^2(\tau)\, d\tau = E. \tag{5.4}$$

This is an average constraint. At the end of the chapter we discuss the incorporation of peak value, or hard constraints, on the instantaneous value of the signal of the form

$$|s(t)| \leqq M. \tag{5.5}$$

We introduce the second constraint, one upon the bandwidth of the signal by using the derivative of the signal. We impose a Gabor, or mean square, bandwidth constraint of the form

$$\int_{T_0}^{T_f} \left(\frac{ds(t)}{dt}\right)^2 dt = \int_{-\infty}^{\infty} \omega^2 |S_s(\omega)|^2 \frac{d\omega}{2\pi} \leqq EB^2. \tag{5.6a}$$

At the end of the chapter we discuss the incorporation of a peak limitation on the relative rate of change of the signal, or constraints of the form

$$\left|\frac{ds(t)}{dt}\right| \leqq E^{1/2}B. \tag{5.6b}$$

While we specifically mention and use these constraints, the methods incorporated can easily be extended to other types of constraints.

We assume that the colored component of the noise is generated by the methods discussed in Chapter 2. By allowing the coefficient matrices to be functions of the transmitted signal, we can model situations where the noise statistics depend upon the transmitted signal. This is a rather general model. We restrict ourselves to the situation where only the observation matrix depends upon $s(t)$. We denote this by the notation $C(t : s(t))$. Several problems of interest can be modeled using this formulation, in particular, the reverberation arising from a Doppler spread environment.*

In summary, we want to choose for our optimization $s(t)$ to maximize d^2 where $g(t)$ is the solution of the integral equation 5.1. Initially we

* Ref. 68, Chap. 4.

impose the contraints of Equations 5.4 and 5.6, and we assume that the colored noise has a state variable method of generation where we allow the possibility that the transmitted signal can affect the generation model via the modulation matrix $\mathbf{C}(t : s(t))$,.

Before developing our approach, we discuss some other proposed approaches to the problem. It is useful here to confine our discussion to the signal-independent channel, or the case where $\mathbf{C}(t)$ is independent of $s(t)$. As first proposed by Middleton and later by Van Trees, an approach that one may want to consider is to formulate the optimization problem in terms of the eigenfunctions and eigenvalues of the homogeneous equation that may be associated with Equation 5.2. If one does this, he finds that the optimal signal is the eigenfunction with the smallest eigenvalue that satisfies Equation 5.6. This approach neglects two important issues. Unless Equation 5.6 is satisfied with equality, we can find better signals. In addition, it neglects discontinuity effects caused by turning the signal on and off at T_0 and T_f, respectively.

A second approach, as proposed by Van Trees, is to apply the calculus of variations while introducing Lagrange multipliers to incorporate the constraints.[69] The resulting integral equation can then be converted to a set of differential equations by using results we derived in Chapter 4. For the particular form of contraints upon the signal that we have initially used, i.e., Equations 5.4 and 5.6, this is undoubtedly the most direct method of the minimization. However, the approach that we use is more general. Many of the results that we develop can be extended to constraints that cannot be readily handled and noise models with the classical calculus of variations.

We assume that the colored component of the noise $y(t)$ is a random process that is generated as we described in Chapter 2. Making this assumption we recognize that we can represent the linear integral Equation 5.1 as a set of differential equations as discussed in the previous chapter. Next, we consider that this set of differential equations can be viewed as a dynamic system with boundary conditions and an input $s(t)$; consequently, the minimum principle of Pontryagin can be used to perform the optimization.[47,3] By using this approach we first find a general solution to the problem; then we consider two examples in order to illustrate the specific techniques involved. This is by no means the only approach to the signal design problem. At the end of the chapter we present a brief survey of some recent work in the area.

* Ref. 67, pp. 302.

5.2 The Application of the Minimum Principle

In this section we develop a state variable formulation for the problem. Using this formulation we apply Pontryagin's minimum principle to find the necessary conditions for the existence of an optimal signal. We then exploit these conditions to find an algorithm for determining the optimal signal.

Since there are several important issues that arise in the course of our derivation, we divide this section into subsections (as listed below):

1. The State Variable Formulation of the Problem
2. The Minimum Principle and the Necessary Conditions
3. The Reduction of the Order of the Equations by $2n$
4. The Asymptotic Solutions

1 *The State Variable Formulation of the Problem*

In order to apply the minimum principle we need to formulate the problem in terms of differential equations, boundary conditions, cost functionals, and a control. First, we need to find a set of differential equations and boundary conditions which relates $g(t)$, the solution of the Fredholm integral equation expressed by the solution of Equation 5.2, to the signal $s(t)$.

We can do this by using the results derived in the previous chapter. Reviewing these results we have shown that $g(t)$, the solution to Equation 5.2, is given by Equation 4.13

$$g(t) = \frac{2}{N_0}(s(t) - \mathbf{C}(t:s(t))\boldsymbol{\xi}(t)), \qquad T_0 \leqq t \leqq T_f. \tag{5.7}$$

The vector function $\boldsymbol{\xi}(t)$ satisfies the differential equations 4.14 and 4.16,

$$\frac{d\boldsymbol{\xi}(t)}{dt} = \mathbf{F}(t)\boldsymbol{\xi}(t) + \mathbf{G}(t)\mathbf{Q}\mathbf{G}^T(t)\boldsymbol{\eta}(t), \qquad T_0 \leqq t \leqq T_f. \tag{5.8}$$

$$\frac{d\boldsymbol{\eta}(t)}{dt} = \mathbf{C}^T(t:s(t))\frac{2}{N_0}\mathbf{C}(t:s(t))\boldsymbol{\xi}(t) - \mathbf{F}^T(t)\boldsymbol{\eta}(t) - \mathbf{C}^T(t:s(t))\frac{2}{N_0}s(t), \qquad T_0 \leqq t \leqq T_f. \tag{5.9}$$

The boundary conditions that specify the solution uniquely are Equations 4.18 and 4.19.

$$\boldsymbol{\xi}(T_0) = \mathbf{P}_0\,\boldsymbol{\eta}(T_0), \tag{5.10}$$

$$\boldsymbol{\eta}(T_f) = \mathbf{0}. \tag{5.11}$$

Consequently, we have the desired result that we can relate $g(t)$ to $s(t)$ by solving two vector differential equations where we have a two-point boundary value condition imposed upon them. Note that although the dependence of the solution upon $s(t)$ is nonlinear when $\mathbf{C}(t : s(t))$ is actually a function of $s(t)$, the equations are still linear in the variables $\boldsymbol{\xi}(t)$ and $\boldsymbol{\eta}(t)$.

Let us now develop the cost functional for the problem. The performance measure of our system is given by Equation 5.2. If we substitute Equation 5.7 in Equation 5.3 and use Equation 5.4, we find that

$$d^2 = \int_{T_0}^{T_f} s(\tau) \frac{2}{N_0} (s(\tau) - \mathbf{C}(\tau : s(\tau))\boldsymbol{\xi}(\tau))\, d\tau$$

$$= \frac{2E}{N_0} - \frac{2}{N_0} \int_{T_0}^{T_f} s(\tau)\mathbf{C}(\tau : s(\tau))\boldsymbol{\xi}(\tau)\, d\tau. \tag{5.12}$$

The first term in Equation 5.12 is the performance when there is just white-noise present. The second term represents the degradation in performance caused by the presence of the colored component of the noise. As in Chapter 4, let us define d_g^2 to be

$$d_g^2 = \frac{2}{N_0} \int_{T_0}^{T_f} s(\tau)\mathbf{C}(\tau : s(\tau))\boldsymbol{\xi}(\tau)\, d\tau, \tag{5.13}$$

and the function $L(\boldsymbol{\xi}(t), s(t))$ to be

$$L(\boldsymbol{\xi}(t), s(t)) = \frac{2}{N_0} s(t)\mathbf{C}(t : s(t))\boldsymbol{\xi}(t), \qquad T_0 \leqq t \leqq T_f, \tag{5.14a}$$

or

$$d_g^2 = \int_{T_0}^{T_f} L(\boldsymbol{\xi}(\tau), s(\tau))\, d\tau. \tag{5.14b}$$

Since the energy E and the white-noise level $N_0/2$ are constants, it is obvious that we can maximize d^2 by minimizing d_g^2.

The state variable formulation requires that the system variables be related by derivative rather than integral operations. Since we are constraining both the signal and its derivative, we cannot use $s(t)$ as the control. Instead, let us define the control function, $v(t)$, to be the derivative of the signal.

$$v(t) = ds(t)/dt. \tag{5.15}$$

Furthermore, we require

$$s(T_0) = s(T_f) = 0. \tag{5.16}$$

Equation 5.16 is a logical requirement. Since we are constraining the derivative of the signal, it is reasonable to require that there be no jump discontinuities (implying singularities in $v(t)$) at the end points of the interval.

We now have all the state equations and boundary conditions that describe the dynamics of the system. The state equations are given by Equations 5.8, 5.9, and 5.15.

$$\frac{d\boldsymbol{\xi}(t)}{dt} = \mathbf{F}(t)\boldsymbol{\xi}(t) + \mathbf{G}(t)\mathbf{Q}\mathbf{G}^T(t)\boldsymbol{\eta}(t), \qquad T_0 \leqq t \leqq T_f, \tag{5.8}$$

$$\frac{d\boldsymbol{\eta}(t)}{dt} = \mathbf{C}^T(t:s(t))\frac{2}{N_0}\mathbf{C}(t:s(t))\boldsymbol{\xi}(t) - \mathbf{F}^T(t)\boldsymbol{\eta}(t) - \mathbf{C}^T(t:s(t))\frac{2}{N_0}s(t), \qquad T_0 \leqq t \leqq T_f, \tag{5.9}$$

$$\frac{ds(t)}{dt} = v(t), \qquad T_0 \leqq t \leqq T_f. \tag{5.15}$$

We have $(2n + 1)$ individual equations. The boundary conditions are given by Equations 5.10, 5.11, and 5.16.

$$\boldsymbol{\xi}(T_0) = \mathbf{P}_0\,\boldsymbol{\eta}(T_0), \tag{5.10}$$

$$\boldsymbol{\eta}(T_f) = \mathbf{0}, \tag{5.11}$$

$$s(T_0) = s(T_f) = 0. \tag{5.16}$$

Notice that there are $(2n + 2)$ individual boundary conditions. Consequently, these conditions cannot be satisfied for an arbitrary $v(t)$.

In order to introduce the energy and bandwidth constraints, we need to augment the state equations artificially by adding the two equations

$$\frac{dx_E(t)}{dt} = \frac{s^2(t)}{2}, \qquad T_0 \leqq t \leqq T_f, \tag{5.17}$$

$$\frac{dx_B(t)}{dt} = \frac{v^2(t)}{2}, \qquad T_0 \leqq t \leqq T_f. \tag{5.18}$$

(We have introduced the factor of 1/2 for a later convenience.) The boundary conditions are

$$x_E(T_0) = x_B(T_0) = 0, \tag{5.19}$$

$$x_E(T_f) = E/2, \tag{5.20}$$

$$x_B(T_f) = EB^2/2. \tag{5.21}$$

It is easy to see that these differential equations and boundary conditions represent the constraints described by Equations 5.4 and 5.6.

With these last results, we have formulated the problem in a form where we can apply the minimum principle.

2 *The Minimum Principle and the Necessary Conditions*

In this section we use Pontryagin's minimum principle to derive the necessary conditions for optimality. Before proceeding, two comments are in order. First, the control function is $v(t)$ not $s(t)$, which is one of the components of the state vector for the system. Secondly, we do not develop much background material on the minimum principle itself. For further information we refer to References 47 and 3.

The Hamiltonian for this system is*

$$
\begin{aligned}
&H(\boldsymbol{\xi}, \boldsymbol{\eta}, s, x_E, x_B, p_0, \mathbf{p}_\xi, \mathbf{p}_\eta, p_s, \lambda_E, \lambda_B, v, t) \\
&\quad = p_0 L(\boldsymbol{\xi}(t), s(t)) + \mathbf{p}_\xi^T(t)\dot{\boldsymbol{\xi}}(t) + \mathbf{p}_\eta^T(t)\dot{\boldsymbol{\eta}}(t) + p_s(t)\dot{s}(t) \\
&\qquad + \lambda_E(t)x_E(t) + \lambda_B(t)x_B(t) \\
&\quad = p_0 \frac{2}{N_0}\mathbf{s}(t)\mathbf{C}(t:s(t))\boldsymbol{\xi}(t) \\
&\qquad + \mathbf{p}_\xi^T(t)(\mathbf{F}(t)\boldsymbol{\xi}(t) + \mathbf{G}(t)\mathbf{Q}\mathbf{G}^T(t)\boldsymbol{\eta}(t)) \\
&\qquad \mathbf{p}_\eta^T(t)\Big(\mathbf{C}^T(t:s(t))\frac{2}{N_0}\mathbf{C}(t:s(t))\boldsymbol{\xi}(t) \\
&\qquad - \mathbf{F}^T(t)\boldsymbol{\eta}(t) - \mathbf{C}^T(t:s(t))\frac{2}{N_0}s(t)\Big) \\
&\qquad p_s(t)v(t) + \lambda_E(t)\frac{s^2(t)}{2} + \lambda_B(t)\frac{v^2(t)}{2}, \qquad T_0 \leqq t \leqq T_f. \quad (5.22)
\end{aligned}
$$

We have denoted the costate vector of the state equation describing the dynamics of the system (Equations 5.8, 5.9, and 5.15) by the variable $\mathbf{p}(t)$. The subscript indicates the corresponding state variable. The costates of the constraint equations are denoted by $\lambda_E(t)$ and $\lambda_B(t)$. The system that we want to optimize may be explicitly time-dependent (nonautonomous), has a fixed time interval, and has boundary conditions at both ends of the time interval.

Let $\boldsymbol{\xi}(t)$, $\boldsymbol{\eta}(t)$, and $s(t)$ be the functions that satisfy the differential equations expressed by Equations 5.8, 5.9, and 5.15, the boundary

* We drop the arguments when there is no specific need for them.

conditions given by Equations 5.10, 5.11, and 5.16, and the constraints of Equations 5.14–5.21 when the control function is $v(t)$. For this problem the minimum principle states: In order that $\hat{v}(t)$ be optimum, it is necessary that there exist a constant p_0 and functions $\hat{\mathbf{p}}_{\xi}(t)$, $\hat{\mathbf{p}}_{\eta}(t)$, $\hat{p}_s(t)$, $\hat{\lambda}_E(t)$, and $\hat{\lambda}_B(t)$ (not all identically zero) such that the following four conditions hold:

Condition a.

$$\dot{\hat{\mathbf{p}}}_{\xi}(t) = -\nabla_{\xi}\hat{H},^{*} \tag{5.23}$$

$$\dot{\hat{\mathbf{p}}}_{\eta}(t) = -\nabla_{\eta}\hat{H}, \tag{5.24}$$

$$\dot{\hat{p}}_s(t) = -\partial\hat{H}/\partial s, \tag{5.25}$$

$$\dot{\hat{\lambda}}_E(t) = -\partial\hat{H}/\partial x_E, \tag{5.26}$$

$$\dot{\hat{\lambda}}_B(t) = \partial\hat{H}/\partial x_B. \tag{5.27}$$

Condition b.

For all t in the interval $[T_0, T_f]$ the function

$H(\hat{\boldsymbol{\xi}}, \boldsymbol{\eta}, \hat{\mathbf{s}}, \hat{x}_E, \hat{x}_B, \hat{\mathbf{p}}_{\xi}, \hat{\mathbf{p}}_{\eta}, \hat{p}_s, \hat{\lambda}_E, \hat{\lambda}_B, v, t)$

is minimized as a function of the variable v.

Condition c.

p_0 is a constant with $p_0 \geqq 0$.

Condition d.

The costate vector is perpendicular to the manifold defined by boundary conditions at each end of the interval.

Let us now examine what each of these assertions implies. If we perform the derivative operations indicated by Equations 5.23 to 5.27 we find†

$$\dot{\hat{\mathbf{p}}}_{\xi}(t) = -p_0 \frac{2}{N_0} \mathbf{C}^T(t:\hat{s}(t))\hat{s}(t) - \mathbf{F}^T(t)\hat{\mathbf{p}}_{\xi}(t)$$
$$- \mathbf{C}(t:\hat{s}(t)) \frac{2}{N_0} \hat{\mathbf{p}}_{\eta}(t), \qquad T_0 \leqq t \leqq T_f; \tag{5.28}$$

* $\hat{H} = H(\boldsymbol{\xi}, \hat{\boldsymbol{\eta}}, \hat{s}, \hat{x}_E, \hat{x}_B, \hat{p}_0, \mathbf{p}_{\xi}, \mathbf{p}_{\eta}, \hat{p}_s, \hat{\lambda}_E, \hat{\lambda}_B, v, t)$.

† $\dfrac{\partial\mathbf{C}(t:s(t))}{\partial s(t)} \triangleq \left[\dfrac{\partial C_i(t:s(t))}{\partial s(t)} \cdots \dfrac{\partial C_n(t:s(t))}{\partial s(t)}\right].$

$$\dot{\hat{\mathbf{p}}}_{\boldsymbol{\eta}}(t) = -\mathbf{G}(t)\mathbf{Q}\mathbf{G}^T(t)\hat{\mathbf{p}}_{\boldsymbol{\xi}}(t) + \mathbf{F}(t)\hat{\mathbf{p}}_{\boldsymbol{\eta}}(t), \qquad T_0 \leqq t \leqq T_f; \tag{5.29}$$

$$\dot{\hat{\mathbf{p}}}_s(t) = -p_0 \frac{2}{N_0} \mathbf{C}(t:\hat{s}(t))\hat{\boldsymbol{\xi}}(t) - p_0 \hat{\boldsymbol{\xi}}^T(t) \frac{\partial \mathbf{C}^T(t:\hat{s}(t))}{\partial \hat{s}(t)} \hat{s}(t)$$
$$- 2\hat{\mathbf{p}}_{\boldsymbol{\eta}}^T(t)\mathbf{C}^T(t\,;\hat{s}(t)) \frac{2}{N_0} \frac{\partial \mathbf{C}^T(t:\hat{s}(t))}{\partial \hat{s}(t)} \boldsymbol{\xi}(t)$$
$$+ \hat{\mathbf{p}}_{\boldsymbol{\eta}}^T(t) \frac{\partial \mathbf{C}^T(t:\hat{s}(t))}{\partial \hat{s}(t)} \frac{2}{N_0} \hat{s}(t) + \frac{2}{N_0} \mathbf{C}(t:\hat{s}(t))\hat{\mathbf{p}}_{\boldsymbol{\eta}}(t) - \hat{\lambda}_E(t)\hat{s}(t),$$
$$T_0 \leqq t \leqq T_f, \tag{5.30}$$

$$\dot{\hat{\lambda}}_E(t) = 0, \tag{5.31}$$

$$\dot{\hat{\lambda}}_B(t) = 0. \tag{5.32}$$

As expected the energy and bandwidth constraint costates are constants (therefore, we drop the time-dependence notation).

Since we have no boundary upon the control region, we can minimize the Hamiltonian versus the variable $v(t)$ by equating the derivative to zero.

$$\left.\frac{\partial \hat{H}}{\partial v}\right|_{v=\hat{v}(t)} = 0 = \hat{p}_s(t) + \hat{\lambda}_B \hat{v}(t), \qquad T_0 \leqq t \leqq T_f. \tag{5.33}$$

Furthermore, for this to be a minimum, we require that

$$\left.\frac{\partial^2 \hat{H}}{\partial v^2}\right|_{v=\hat{v}(t)} \geqq 0, \tag{5.34a}$$

equivalently, from 5.33,

$$\hat{\lambda}_B \geqq 0. \tag{5.34b}$$

In general, we can show that $\hat{\lambda}_B > 0$. Then we can solve Equation 5.33 for $\hat{v}(t)$. This yields

$$\hat{v}(t) = -\hat{p}_s(t)/\hat{\lambda}_B, \qquad T_0 \leqq t \leqq T_f. \tag{5.35}$$

When we substitute Equation 5.34 in 5.15, we can eliminate the variable $\hat{v}(t)$. Consequently we obtain a set of differential equations in the variables $\hat{\boldsymbol{\xi}}(t)$. $\hat{\boldsymbol{\eta}}(t)$, $\hat{s}(t)$, $\hat{\mathbf{p}}_{\boldsymbol{\xi}}(t)$, $\hat{\mathbf{p}}_{\boldsymbol{\eta}}(t)$, and $\hat{p}_s(t)$ that the optimum signal must satisfy. These are summarized below.

$$\dot{\hat{\boldsymbol{\xi}}}(t) = \mathbf{F}(t)\hat{\boldsymbol{\xi}}(t) + \mathbf{G}(t)\mathbf{Q}\mathbf{G}^T(t)\hat{\boldsymbol{\eta}}(t), \tag{5.36a}$$

$$\dot{\hat{\boldsymbol{\eta}}}(t) = \mathbf{C}^T(t : \hat{s}(t))\frac{2}{N_0}\mathbf{C}(t : \hat{\mathbf{s}}(t))\hat{\boldsymbol{\xi}}(t) - \mathbf{F}^T(t)\hat{\boldsymbol{\eta}}(t) - \mathbf{C}^T(t : \hat{s}(t))\frac{2}{N_0}\hat{s}(t), \tag{5.36b}$$

$$\dot{\hat{s}}(t) = -(1/\hat{\lambda}_B)\hat{p}_s(t), \tag{5.36c}$$

$$\begin{aligned}\dot{\hat{\mathbf{p}}}_{\xi}(t) = &- p_0\,\mathbf{C}^T(t : \hat{s}(t))\hat{s}(t) - \mathbf{F}^T(t)\hat{\mathbf{p}}_{\xi}(t)\\ &- \mathbf{C}^T(t : \hat{s}(t))\frac{2}{N_0}\mathbf{C}(t : \hat{s}(t))\hat{\mathbf{p}}_{\eta}(t),\end{aligned} \tag{5.36d}$$

$$\dot{\hat{\mathbf{p}}}_{\eta}(t) = -\mathbf{G}(t)\mathbf{Q}\mathbf{G}^T(t)\hat{\mathbf{p}}_{\xi}(t) + \mathbf{F}(t)\hat{\mathbf{p}}_{\eta}(t), \tag{5.36e}$$

$$\begin{aligned}\dot{\hat{\mathbf{p}}}_s(t) = &-p_0\frac{2}{N_0}\mathbf{C}(t : \hat{s}(t))\hat{\boldsymbol{\xi}}(t) - p_0\,\hat{\boldsymbol{\xi}}^T(t)\frac{\partial \mathbf{C}^T(t : \hat{s}(t))}{\partial \hat{s}(t)}\hat{s}(t)\\ &- 2\hat{\mathbf{p}}_{\eta}^T(t)\mathbf{C}^T(t : \hat{s}(t))\frac{2}{N_0}\frac{\partial \mathbf{C}(t : \hat{s}(t))}{\partial \hat{s}(t)}\hat{\boldsymbol{\xi}}(t)\\ &+ \hat{\mathbf{p}}_{\eta}^T(t)\frac{\partial \mathbf{C}^T(t : \hat{s}(t))}{\partial \hat{s}(t)}\frac{2}{N_0}\hat{s}(t) + \frac{2}{N_0}\mathbf{C}(t : \hat{s}(t))\hat{\mathbf{p}}_{\eta}(t) - \hat{\lambda}_E\hat{s}(t),\\ &T_0 \leqq t \leqq T_f.\end{aligned} \tag{5.36f}$$

(From now on, we drop the caret notation and assume that we refer to the optimal solution.)

If, in assertion (c) of the minimum principle, the constant p_0 is identically zero, then we have what we call an asymptotic case. We return to this case later; however, let us for the interim set p_0 equal to unity. Since the costate equations are linear in the costate variables this entails no loss of generality.

Let us consider the boundary, or transversality, conditions implied by assertion (d). In order to be perpendicular to the manifold of Equation 5.10, the variables $\mathbf{P}_{\xi}\ (T_0)$ and $\mathbf{p}_{\eta}(T_0)$ must satisfy

$$\mathbf{p}_{\eta}(T_0) = -\mathbf{P}_0\,\mathbf{p}_{\xi}(T_0). \tag{5.37}$$

Equation 5.16 implies that $p_s(T_0)$ is unspecified. At the final, or end point, time we have

$$\mathbf{p}_{\xi}(T_f) = \mathbf{0}, \tag{5.38}$$

and $\mathbf{p}_\eta(T_f)$ and $p_s(T_f)$ are unspecified. We also have that λ_E and λ_B are unspecified constants ($\lambda_B \geqq 0$). We also have the boundary conditions given by Equations 5.10, 5.11, and 5.16. Therefore, we have a total of $4n + 2$ boundary conditions. In addition, we should notice that we do not have to find the control $v(t)$ in order to find $s(t)$, although we may easily deduce it from Equation 5.35.

3 The Reduction of the Order of the Equations by 2n

We are now in a position to show how the assertions of the minimum principle may be used to find the candidates for the optimal signal. However, before proceeding we derive a result that significantly simplifies the solution method. We prove that, in general,

$$\boldsymbol{\xi}(t) = -\mathbf{p}_\eta(t), \qquad T_0 \leqq t \leqq T_f, \tag{5.39}$$

$$\boldsymbol{\eta}(t) = \mathbf{p}_\xi(t), \qquad T_0 \leqq t \leqq T_f. \tag{5.40}$$

This reduction was suggested by the variational approach of Van Trees.[69] We point out, however, that our derivation is independent of the type of constraint imposed upon the signal; i.e., it depends only upon the differential equations for $\boldsymbol{\xi}(t)$ and $\boldsymbol{\eta}(t)$.

Let us define two vectors $\boldsymbol{\varepsilon}_1(t)$ and $\boldsymbol{\varepsilon}_2(t)$ as

$$\boldsymbol{\varepsilon}_1(t) = \boldsymbol{\xi}(t) + \mathbf{p}_\eta(t), \qquad T_0 \leqq t \leqq T_f, \tag{5.41}$$

$$\boldsymbol{\varepsilon}_2(t) = \boldsymbol{\eta}(t) - \mathbf{p}_\xi(t), \qquad T_0 \leqq t \leqq T_f. \tag{5.42}$$

If we differentiate these two equations and substitute Equations 5.8, 5.9, 5.28, and 5.29 (with p_0 equal to 1), we find

$$\begin{aligned}\dot{\boldsymbol{\varepsilon}}_1(t) &= \mathbf{F}(t)\boldsymbol{\xi}(t) + \mathbf{G}(t)\mathbf{Q}\mathbf{G}^T(t)\boldsymbol{\eta}(t) - \mathbf{G}(t)\mathbf{Q}\mathbf{G}^T(t)\mathbf{p}_\xi(t) + \mathbf{F}(t)\mathbf{p}_\eta(t)\\ &= \mathbf{F}(t)\boldsymbol{\varepsilon}_1(t) + \mathbf{G}(t)\mathbf{Q}\mathbf{G}^T(t)\boldsymbol{\varepsilon}_2(t), \qquad T_0 \leqq t \leqq T_f,\end{aligned} \tag{5.43}$$

$$\begin{aligned}\dot{\boldsymbol{\varepsilon}}_2(t) &= \mathbf{C}^T(t:s(t))\frac{2}{N_0}\mathbf{C}(t:s(t))\boldsymbol{\xi}(t) - \mathbf{F}^T(t)\boldsymbol{\eta}(t) - \mathbf{C}^T(t)\frac{2}{N_0}s(t)\\ &\quad + \mathbf{C}^T(t:s(t))\frac{2}{N_0}\mathbf{C}(t:s(t))\hat{\mathbf{p}}_\eta(t) + \mathbf{F}^T(t)\hat{\mathbf{p}}_\xi(t) + \mathbf{C}^T(t)\frac{2}{N_0}s(t)\\ &= \mathbf{C}^T(t:s(t))\frac{2}{N_0}\mathbf{C}(t:s(t))\boldsymbol{\varepsilon}_1(t) - \mathbf{F}^T(t)\boldsymbol{\varepsilon}_2(t), \qquad T_0 \leqq t \leqq T_f.\end{aligned} \tag{5.44}$$

The boundary conditions that the solution to these differential equations satisfy may be found by using Equations 5.8, 5.9, 5.35, and 5.38. They are

$$\boldsymbol{\varepsilon}_1(T_0) = \mathbf{P}_0\,\boldsymbol{\eta}(T_0) - \mathbf{P}_0\,\mathbf{p}_\xi(T_0) = \mathbf{P}_0\,\boldsymbol{\varepsilon}_2(T_0), \tag{5.45}$$

$$\boldsymbol{\varepsilon}_2(T_f) = \boldsymbol{\eta}(T_f) - \mathbf{p}_\xi(T_f) = \mathbf{0}. \tag{5.46}$$

Consequently, Equations 5.43 to 5.46 specify two vector linear differential equations with a two-point boundary value condition. However, these equations are just those that specify the eigenvalues and eigenfunctions for the homogeneous Fredholm integral equation as shown in Chapter 3, Equations 3.23 to 3.25. We have shown that, in order to have a nontrivial solution to this problem, we require that

$$-N_0/2 = \lambda_i > 0, \tag{5.47}$$

where λ_i is an eigenvalue of a Karhunen-Loeve expansion of the colored noise. Clearly, this is impossible since $\lambda_i \geqq 0$. Consequently, the only solution is the trivial one, i.e., $\boldsymbol{\varepsilon}_1(t) = \boldsymbol{\varepsilon}_2(t) = \mathbf{0}$, which proves the assertion of Equations 5.39 and 5.40.

When we make the substitutions implied by Equation 5.39, we find that the problem can be appreciably simplified. We now have that the necessary conditions for an optimal signal are given by

$$\dot{\boldsymbol{\xi}}(t) = \mathbf{F}(t)\boldsymbol{\xi}(t) + \mathbf{G}(t)\mathbf{Q}\mathbf{G}^T(t)\boldsymbol{\eta}(t), \tag{5.48a}$$

$$\dot{\boldsymbol{\eta}}(t) = \mathbf{C}^T(t:s(t))\frac{2}{N_0}\mathbf{C}(t:s(t))\boldsymbol{\xi}(t) - \mathbf{F}^T(t)\boldsymbol{\eta}(t) - \mathbf{C}^T(t:s(t))\frac{2}{N_0}s(t), \tag{5.48b}$$

$$\dot{s}(t) = -(1/\lambda_B)p_s(t), \tag{5.48c}$$

$$\dot{p}_s(t) = -\frac{4}{N_0}\mathbf{C}(t:s(t))\boldsymbol{\xi}(t) + \frac{4}{N_0}(\boldsymbol{\xi}^T(t)\mathbf{C}^T(t:s(t)))\frac{\partial\mathbf{C}(t:s(t))}{\partial s(t)}\boldsymbol{\xi}(t) - \lambda_E s(t), \tag{5.48d}$$

with boundary conditions

$$\mathbf{P}_0\boldsymbol{\eta}(T_0) = \boldsymbol{\xi}(T_0), \tag{5.49a}$$

$$\boldsymbol{\eta}(T_f) = \mathbf{0}, \tag{5.49b}$$

$$s(T_0) = s(T_f) = 0. \tag{5.49c}$$

4 The Asymptotic Solutions

In our earlier discussion, we deferred the discussion of the situation when p_0 equals zero. Since the solutions that result are useful in the analysis of a particular problem, it is worthwhile to examine them before proceeding with the discussion of the algorithm of our design procedure. We term the solutions that satisfy the necessary conditions of the minimum principle when p_0 is zero the asymptotic solutions. (They are often called pathological solutions.) In order to test for their existence, we set p_0 equal to zero in Equation 5.36 and examine the differential equations for $\mathbf{p}_\xi(t)$ and $\mathbf{p}_\eta(t)$. If we write these equations in augmented vector form, we obtain the following homogeneous equation:

$$\frac{d}{dt}\begin{bmatrix}\mathbf{p}_\xi(t)\\ \mathbf{p}_\eta(t)\end{bmatrix}=\begin{bmatrix}-\mathbf{F}^T(t) & -\mathbf{C}^T(t:s(t))\dfrac{2}{N_0}\mathbf{C}(t:s(t))\\ -\mathbf{G}(t)\mathbf{Q}\mathbf{G}^T(t) & \mathbf{F}(t)\end{bmatrix}\begin{bmatrix}\mathbf{p}_\xi(t)\\ \mathbf{p}_\eta(t)\end{bmatrix},$$

$$T_0 \leqq t \leqq T_f. \quad (5.50)$$

The appropriate boundary conditions are specified by Equations 5.37 and 5.38.

From Chapter 3, Equation 3.14,

$$\begin{bmatrix}-\mathbf{F}^T(t) & -\mathbf{C}^T(t:s(t))\dfrac{2}{N_0}\mathbf{C}(t:s(t))\\ -\mathbf{G}(t)\mathbf{Q}\mathbf{G}^T(t) & \mathbf{F}(t)\end{bmatrix}$$

$$= -\mathbf{W}^T(t:N_0/2). \quad (5.51)$$

Let us define the transition matrix associated with Equation 5.51 to be $\mathbf{\Phi}(t, \tau : -N_0/2)$. We note that $\mathbf{\Phi}(t, \tau : -N_0/2)$ is related to the transition matrix $\mathbf{\Psi}(t, \tau : -N_0/2)$ associated with $\mathbf{W}(t : N_0/2)$

$$\mathbf{\Phi}(t, \tau : -N_0/2) = \mathbf{\Psi}^T(\tau, t : -N_0/2). \quad (5.52)$$

Any solution of Equation 5.50 may be found in terms of $\mathbf{\Phi}(t, \tau : -N_0/2)$. In order to find the solution to Equation 5.50 that satisfies the boundary conditions, we partition $\mathbf{\Phi}(t, \tau : -N_0/2)$ into four $n \times n$ submatrices,

$$\mathbf{\Phi}(t:\tau:N_0/2) = \begin{bmatrix}\mathbf{\Psi}_{\xi\xi}^T(\tau, t: -N_0/2) & \mathbf{\Psi}_{\eta\xi}^T(\tau, t: -N_0/2)\\ \mathbf{\Psi}_{\xi\eta}^T(\tau, t: -N_0/2) & \mathbf{\Psi}_{\eta\eta}^T(\tau, t: -N_0/2)\end{bmatrix}.$$

$$(5.53)$$

If we incorporate the boundary condition specified by Equation 5.38 that the solution to Equation 5.51 is

$$\begin{bmatrix} \mathbf{p}_\xi(t) \\ \mathbf{p}_\eta(t) \end{bmatrix} = \begin{bmatrix} \mathbf{\Psi}^T_{\eta\xi}(T_f, t : -N_0/2) \\ \mathbf{\Psi}^T_{\eta\eta}(T_f, t : -N_0/2) \end{bmatrix} \mathbf{p}_\eta(T_f). \tag{5.54}$$

The initial condition specified by Equation 5.36 requires

$$\begin{aligned} \mathbf{p}_\eta(T_0) &= \mathbf{\Psi}^T_{\eta\eta}(T_f, T_0 : -N_0/2)\mathbf{p}_\eta(T_f) \\ &= -P_0\,\mathbf{p}_\xi(T_0) = -\mathbf{P}_0\mathbf{\Psi}^T_{\eta\xi}(T_f, T_0 : -N_0/2)\mathbf{p}_\eta(T_f), \end{aligned} \tag{5.55}$$

or

$$\mathbf{0} = [\mathbf{\Psi}^T_{\eta\eta}(T_f, T_0 : -N_0/2) + \mathbf{P}_0\mathbf{\Psi}^T_{\eta\xi}(T_f, T_0 : -N_0/2)]\mathbf{p}_\eta(T_f). \tag{5.56}$$

The only way that Equation 5.56 can have a nontrivial solution is for the determinant of the matrix enclosed by brackets in Equation 5.56 to vanish.

$$\det[\mathbf{\Psi}^T_{\eta\eta}(T_f, T_0 : -N_0/2) + \mathbf{P}_0\mathbf{\Psi}^T_{\eta\xi}(T_f, T_0 : -N_0/2)] = 0. \tag{5.57}$$

If we transpose the matrix in Equation 5.57 (this does not change the determinant value), we find that we have the test for an eigenvalue that we developed in Chapter 3. There we showed that the only way for this determinant to vanish is for

$$-N_0/2 = \lambda_i > 0,$$

where λ_i is a eigenvalue associated with the Karhunen-Loeve expansion colored noise process. Clearly, this is impossible. Consequently, the only solution is the trivial one; i.e.,

$$\mathbf{p}_\xi(t) = \mathbf{p}_\eta(t) = \mathbf{0}, \qquad T_0 \leqq t \leqq T_f. \tag{5.58}$$

If we substitute Equation 5.58 in Equation 5.36, we find that the differential equations for $s(t)$ and $p_s(t)$ are

$$\dot{s}(t) = -\frac{1}{\lambda_B} p_s(t), \qquad T_0 \leqq t \leqq T_f, \tag{5.59}$$

$$\dot{p}_s(t) = -\lambda_E s(t), \qquad T_0 \leqq t \leqq T_f. \tag{5.60}$$

The only solution to these equations that satisfies the boundary conditions specified by Equation 5.16 is

$$s(t) = \sqrt{\frac{2E}{T}} \sin\left(n\pi \frac{t - T_0}{T_f - T_0}\right), \qquad T_0 \leqq t \leqq T_f. \tag{5.61}$$

with

$$-\frac{\lambda_E}{\lambda_B} = \left(\frac{n\pi}{T_f - T_0}\right)^2, \qquad \lambda_E < 0. \tag{5.62}$$

Several comments are in order. (1) The bandwidth of these signals is easily shown to be $n\pi/T$. (2) We have not violated the assertion (a) in our application of the minimum principle since $p_s(t)$ is nonzero. (3) We did not require our system to be time invariant, nor did we specify the dimension of the system. (4) We point out that these solutions are only for this particular constraint set. Consequently, these solutions, which are probably one of the most practical to transmit, exist for all types of colored noise that fit within our model.

These equations specify the necessary conditions for an optimal signal when it is imbedded in $\mathbf{C}(t : s(t))$ in an arbitrary manner. One can generalize the problem by imbedding the signal in one of the other coefficient matrices of the state variable description of the noise. One can also introduce other constraint forms, and we do this briefly at the end of the specific cases that we develop. We now consider two special cases. The first is the case when the noise is signal independent; i.e., the observation matrix is not a function of the transmitted signal. This is an important situation, and it has been discussed extensively.[45,67] Because of the linearity of the resulting problem we can develop a complete solution method and illustrate specific examples. The second case of interest is when $\mathbf{C}(t : s(t))$ has the form $s(t)\mathbf{C}_0(t)$; i.e., the noise is multiplicative, and it corresponds to a Doppler spread channel. Here finding the necessary conditions is straightforward; however, determining a solution method is not, essentially because of the nonlinear aspects of the problems.

5.3 Optimal Signal Design for Additive Signal-Independent Noise Channels*

In this section we consider the optimal signal design problem when the characteristics of the channel noise do not depend upon $s(t)$, or when

$$\mathbf{C}(t : s(t)) = \mathbf{C}(t). \tag{5.63}$$

First we use the necessary conditions derived above to determine a transcendental equation relating λ_E and λ_B. In general, this equation has a multimodal solution. We therefore derive a relation that specifies globally which solution is desired. After discussing some asymptotic

* This material has appeared in Ref. 11.

solutions, we then suggest a signal design algorithm for actually obtaining the optimal signals and their resulting performance. We illustrate this with two examples, and finally we discuss some generalizations.

The derivation and the final result for the transcendental equation relating λ_E and λ_B are similar to those used in Chapter 3 to find the eigenvalues of the homogeneous Fredholm equation. The most important distinction is that this equation is in terms of two parameters, λ_E and λ_B, whereas we had only one previously. Once we satisfy this equation, we can generate a signal that is a candidate for the optimum solution.

Because of the form expressed by Equation 5.63, Equations 5.48 are linear in the variables $\boldsymbol{\xi}(t)$, $\boldsymbol{\eta}(t)$, $s(t)$, and $p_s(t)$. Consequently, we can write them in a $(2n + 2)$-dimensional matrix form

$$\frac{d}{dt}\begin{bmatrix} \boldsymbol{\xi}(t) \\ \boldsymbol{\eta}(t) \\ s(t) \\ p_s(t) \end{bmatrix} = \begin{bmatrix} \mathbf{F}(t) & \mathbf{G}(t)\mathbf{Q}\mathbf{G}^T(t) & \mathbf{0} & \mathbf{0} \\ \mathbf{C}^T(t)\dfrac{2}{N_0}\mathbf{C}(t) & -\mathbf{F}^T(t) & -\mathbf{C}^T(t)\dfrac{2}{N_0} & 0 \\ \mathbf{0} & \mathbf{0} & 0 & -\dfrac{1}{\lambda_B} \\ -\dfrac{4}{N_0}\mathbf{C}(t) & \mathbf{0} & -\lambda_E & \mathbf{0} \end{bmatrix}\begin{bmatrix} \boldsymbol{\xi}(t) \\ \boldsymbol{\eta}(t) \\ s(t) \\ p_s(t) \end{bmatrix}$$

$$T_0 \leqq t \leqq T_f. \qquad (5.64)$$

The boundary conditions are given by Equations 5.49a, 5.49b, and 5.49c. These conditions specify $2n + 2$ boundary conditions that must be satisfied for an optimum to exist.

Since Equation 5.64 is a homogeneous linear equation, we may not, in general, have a nontrivial solution. In order to find where we may obtain a nontrivial solution, let us define the transition matrix associated with Equation 5.64 to be $\boldsymbol{\chi}(t, T_0 : \lambda_E, \lambda_B)$. We emphasize the dependence of $\boldsymbol{\chi}$ upon λ_E and λ_B by including them as arguments.

Since Equation 5.64 is linear, we can determine any solution to it in terms of this transition matrix. If we use the boundary conditions specified by Equation 5.49, we find that any solution that satisfies the initial conditions may be written in the form

$$\begin{bmatrix} \boldsymbol{\xi}(t) \\ \boldsymbol{\eta}(t) \\ s(t) \\ p_s(t) \end{bmatrix} = \boldsymbol{\chi}(t, T_0; \lambda_E, \lambda_B)\begin{bmatrix} \mathbf{P}_0\,\boldsymbol{\eta}(T_0) \\ \boldsymbol{\eta}(T_0) \\ 0 \\ p_s(T_0) \end{bmatrix} \qquad T_0 \leqq t \leqq T_f. \qquad (5.65)$$

The final boundary condition requires that $\boldsymbol{\eta}(T_f)$ and $s(T_f)$ both be zero. In order to see what this implies, let us partition this transition matrix as follows (we drop the arguments temporarily):

$$\boldsymbol{\chi} = \begin{bmatrix} \chi_{\xi\xi} & \chi_{\xi\eta} & \chi_{\xi_s} & \chi_{\xi p_s} \\ \chi_{\eta\xi} & \chi_{\eta\eta} & \chi_{\eta_s} & \chi_{\eta p_s} \\ \chi_{s\xi} & \chi_{s\eta} & \chi_{ss} & \chi_{sp_s} \\ \chi_{p_s\xi} & \chi_{p_s\eta} & \chi_{p_s s} & \chi_{p_s p_s} \end{bmatrix}. \tag{5.66}$$

By substituting Equation 5.66 in Equation 5.65, we find that, in terms of these partitions, the requirement that $\boldsymbol{\eta}(T_f)$ vanish implies

$$\mathbf{0} = [\chi_{\eta\xi}(T_f, T_0 : \lambda_E, \lambda_B)\mathbf{P}_0 + \chi_{\eta\eta}(T_f, T_0 : \lambda_E, \lambda_B)]\boldsymbol{\eta}(T_0) \\ + \chi_{\eta p_s}(T_f, T_0 : \lambda_E, \lambda_B)]p_s(T_0). \tag{5.67}$$

Similarly, we find that $s(T_f)$ being zero requires

$$0 = [\chi_{s\xi}(T_f, T_0 : \lambda_E, \lambda_B)\mathbf{P}_0 + \chi_{s\eta}(T_f, T_0 : \lambda_E, \lambda_B)]\boldsymbol{\eta}(T_0) \\ + \chi_{sp_s}(T_f, T_0 : \lambda_E, \lambda_B)p_s(T_0). \tag{5.68}$$

We can write Equations 5.67 and 5.68 more concisely in matrix-vector form

$$\mathbf{0} = \begin{bmatrix} \chi_{\eta\xi}(T_f, T_0 : \lambda_E, \lambda_B)\mathbf{P}_0 & \chi_{\eta p_s}(T_f, T_0 : \lambda_E, \lambda_B) \\ \quad + \chi_{\eta\eta}(T_f, T_0 : \lambda_E, \lambda_B) & \\ \chi_{s\xi}(T_f, T_0 : \lambda_E, \lambda_B)\mathbf{P}_0 & \chi_{sp_s}(T_f, T_0 : \lambda_E, \lambda_B) \\ \quad + \chi_{ps_s}(T_f, T_0 : \lambda_E, \lambda_B) & \end{bmatrix} \begin{bmatrix} \boldsymbol{\eta}(T_0) \\ \\ p_s(T_0) \\ \end{bmatrix} \tag{5.69}$$

or by defining the matrix in Equation 5.69 to be $\mathbf{D}(\lambda_E, \lambda_B)$, we have

$$\mathbf{0} = \mathbf{D}(\lambda_E, \lambda_B)\begin{bmatrix} \boldsymbol{\eta}(T_0 \\ p_s(T_0) \end{bmatrix}. \tag{5.70}$$

Equation 5.69 specifies a set of $n + 1$ linear homogeneous algebraic equations. The only way that this set of equations can have a nontrivial solution is for the determinant of the matrix $\mathbf{D}(\lambda_E, \lambda_B)$ to be identically zero. Consequently, the test for candidates for the optimal signal is to find those values of λ_E and λ_B (>0) such that

$$\det[\mathbf{D}(\lambda_E, \lambda_B)] = 0. \tag{5.71}$$

Once Equation 5.71 is satisfied, we can find a nonzero solution to Equation 5.70 up to a multiplicative constant. Knowing $\boldsymbol{\eta}(T_0)$ and

$p_s(T_0)$ allows us to determine the candidate signal(s), $s_{\lambda_E,\lambda_B}(t)$, for the particular values of λ_E and λ_B that satisfy Equation 5.71. The multiplicative constant may be determined by applying the energy constraint of Equation 5.4, i.e.,

$$\int_{T_0}^{T_f} s_{\lambda_E,\lambda_B}^2(\tau)\,d\tau = E. \tag{5.72}$$

By using Equation 5.35 we can determine the bandwidth of the signal. We have that

$$B^2 = \frac{1}{E}\int_{T_0}^{T_f}\left(\frac{ds_{\lambda_E,\lambda_B}(\tau)}{d\tau}\right)^2 d\tau = \frac{1}{E\lambda_B^2}\int_{T_0}^{T_f} p_{s\lambda_E,\lambda_B}^2(\tau)\,d\tau. \tag{5.73}$$

In order to satisfy Equation 5.71, we require that only the rank of $\mathbf{D}(\lambda_E, \lambda_B)$ be less than or equal to n, the dimension of $\boldsymbol{\xi}(t)$ and $\boldsymbol{\eta}(t)$. The case when this rank is less than n presents an important aspect of this optimization. For convenience, let us define

$$n_D = n + 1 - \text{Rank}[\mathbf{D}(\lambda_E, \lambda_B)], \tag{5.74}$$

where n_D specifies the number of linearly independent solutions to Equation 5.70 which we may obtain for the given values of λ_E and λ_B. These solutions in turn specify n_D functions, $v(t)$, that satisfy the necessary conditions for optimality given by the minimum principle.

We see that, because of the linearity of 5.65, any linear combination of these functions having the same values of λ_E and λ_B also satisfies the necessary conditions given by the minimum principle. Consequently, any time we find n_D is greater than 1, we must consider these linear combinations when checking to see which candidate is indeed optimum. Of course, these candidates are subject to the same constraints as any other, i.e., the energy and bandwidth constraints given by Equations 5.72 and 5.73.

In general, for a given value of λ_E, Equation 5.71 may have one or an infinite number of solutions. Each of these solutions in turn specifies a candidate for the optimal signal. Consequently, it is desirable to be able to relate the parameters λ_E and λ_B to the constraint values E and B and to the performance. In this way we can specify which candidate is optimum globally.

We have from Equation 5.64 that

$$\dot{s}(t) = -\frac{1}{\lambda_B}\,p_s(t). \tag{5.75}$$

Differentiating this, multiplying by $s(t)$, and integrating by parts over the interval $[T_0, T_f]$ yields

$$-\int_{T_0}^{T_f} \frac{1}{\lambda_B} \dot{p}_s(t)s(t)\,dt = \int_{T_0}^{T_f} \ddot{s}(t)s(t)\,dt = -\int_{T0}^{T_f} (\dot{s}(t))^2\,dt = -EB^2, \tag{5.76}$$

where we have applied the boundary conditions upon $s(t)$ and used Equation 5.6. If we use Equation 5.64 and substitute for $\dot{p}_s(t)$, we obtain

$$\frac{1}{\lambda_B}\int_{T_0}^{T_f}\left[-\frac{4}{N_0}\mathbf{C}(t)\xi(t) - \lambda_E s(t)\right]s(t)\,dt$$

$$= -\frac{1}{\lambda_B}[2d_g^2 + \lambda_E E] = EB^2, \tag{5.77}$$

where we have used Equations 5.4 and 5.13 for the energy E and degradation d_g^2, respectively. Consequently, we have

$$d_g^2 = -\frac{E}{2}(\lambda_E + \lambda_B B^2). \tag{5.78}$$

We can observe two useful results here. First, E, B^2, d_g^2, and λ_B are all nonnegative; therefore, λ_E must be negative with

$$-\lambda_E > \lambda_B B^2. \tag{5.79}$$

Second, we have that, for a given value of λ_E, the largest solution for λ_B minimizes d_g^2, the degradation. Consequently, we need to examine only one locus of solutions of Equation 5.71 for the globally optimum signal. This result is analogous to the smallest eigenvalue optimization procedure when there is no bandwidth constraint imposed.

We have now derived the results necessary for designing optimal signals for this channel. It is useful to summarize these results in the context of an algorithm which when implemented on a digital computer yields the optimal signals and their performances.

Ideally, we want to be able to find the optimum signal for given values of E and B. Although this is certainly possible, it is far more efficient to solve a particular problem where we let B be a parameter and then select the specific value in which we are interested (the energy is normalized to unity). The reason for this approach will become apparent when we consider some specific examples.

The minimum principle has provided us with a set of necessary conditions from which we found a test (Equation 5.71) for an optimum signal. The result of this test is that we essentially have an eigenvalue problem in two dimensions. The most difficult aspect of the problem becomes finding the particular values of λ_E and λ_B that both satisfy this test and correspond to a signal with the desired bandwidth. The algorithm that we suggest here is simply a systematic method of approaching this aspect of the problem.

The algorithm has several steps. First, we outline it. Then in the next section we discuss it in the context of two examples.

a. First we find the loci points in the λ_E-λ_B plane that satisfy 5.71. Equations 5.34b and 5.79 imply that we can confine our attention to the quadrant where $\lambda_E < 0$ and $\lambda_B > 0$. Since we are interested in only the globally optimum signal, Equation 5.78 also implies that for a specified value of λ_E we need only to find the largest value of λ_B satisfying 5.58. Plotting λ_B versus λ_E generates a locus of points. We need to be careful with the situation where $\mathbf{D}(\lambda_E, \lambda_B)$ has rank less than n, as we indicated earlier. If we plot the loci corresponding to nonglobally optimum solutions, we find that this situation occurs when the loci cross. It is useful, therefore, to plot the loci corresponding to the largest $n + 1$ solutions of Equation 5.58 rather than just the largest one. In this way, we can conveniently determine the loci crossings and those points where $\mathbf{D}(\lambda_E, \lambda_B)$ has rank less than n. We also note that this requires that we have an effective procedure for calculating transition matrices, as discussed in Chapter 3. In general, one can simply numerically integrate the differential equations that specify the transition matrix. However, if the matrices that describe the generation of the channel noise (F, G, Q, C) are constants, then we can use the matrix exponential, i.e.,

$$\boldsymbol{\chi}(t, T_0 : \lambda_E, \lambda_B) = e^{\mathbf{Z}(\lambda_E, \lambda_B)(t - T_0)}, \tag{5.80}$$

where $\mathbf{Z}(\lambda_E, \lambda_B)$ is the coefficient matrix of Equation 5.64.

b. For a particular point on these loci, solve Equation 5.70 for $\boldsymbol{\eta}(T_0)$ and $p_s(T_0)$. Then use Equation 5.65 to determine the signals $s_{\lambda_E, \lambda_B}(t)$, $p_{s_{\lambda_E, \lambda_B}}(t)$, and $\xi_{\lambda_E, \lambda_B}(t)$

c. Since the performance is linearly related to the energy, normalize these signals such that $s_{\lambda_E, \lambda_B}(t)$ has unit energy.

d. Calculate the bandwidth and performance of the normalized signals as specified by Equations 5.73 and 5.14, respectively.

e. Repeat parts b, c, and d at appropriate intervals along these loci in the λ_E-λ_B plane. (The interval should be small enough so that the bandwidth and performance as calculated in part d vary in a reasonably continuous manner.) As we move along a particular locus in the λ_E-λ_B plane, plot the degradation versus bandwidth in a second plane, a d_g^2-B^2 plane.

f. As mentioned earlier, we need to pay special attention to the case when Equation 5.70 has more than one solution. This situation corresponds to a crossing of two or more loci in the λ_E-λ_B plane. In this case find the solutions and plot the locus produced in the d_g^2-B^2 plane by linearly combining the different signals. Probably the most convenient means of doing this is to use the Fredholm integral equation technique discussed in Chapter 4.

g. As B^2 is increased, one eventually approaches the "matched filter" receiver and the degradation due to the colored components of the noise is not significant.

We now illustrate our design algorithm with two examples. We first consider the case when the colored noise has a one-pole spectrum as described in Chapter 2 (Equations 2.25, 2.26) and then for the two-pole spectrum, which is also described in Chapter 2 (Equations 2.28, 2.29).

Example 5.1 Signal Design for Channels with a First-Order Noise Spectrum

Let us assume that the colored noise component has a first-order spectrum, which has the form

$$S_y(\omega) = \frac{2kP}{\omega^2 + k^2}.$$

The equations describing the generation of this process are given by Equations 2.25, 2.26. We also set $T_0 = 0$ and $T_f = T$.

In order to set up the test for the optimum signal, we need to find the coefficient matrix in Equation 5.64. If we substitute the coefficients of Equations 2.25 and 2.26 into Equation 5.64, we obtain

$$\frac{d}{dt}\begin{bmatrix} \xi(t) \\ \eta(t) \\ s(t) \\ p_s(t) \end{bmatrix} = \begin{bmatrix} -k & 2kP & 0 & 0 \\ \frac{2}{N_0} & k & -\frac{2}{N_0} & 0 \\ 0 & 0 & 0 & -\frac{1}{\lambda_B} \\ -\frac{4}{N_0} & 0 & -\lambda_E & 0 \end{bmatrix} \begin{bmatrix} \xi(t) \\ \eta(t) \\ s(t) \\ p_s(t) \end{bmatrix}. \tag{5.81}$$

From Equations 5.49 and 2.25 and boundary conditions are

$$\xi(0) = P\eta(0), \tag{5.82a}$$

$$\eta(T) = 0, \tag{5.82b}$$

$$s(0) = s(T) = 0. \tag{5.82c}$$

Step (a) in our algorithm requires us to find the loci in λ_E-λ_B plane that satisfy Equation 5.71. To do this, we need to calculate the transition matrix (4×4) of Equation 5.81, $\chi(T; 0 : \lambda_E, \lambda_B)$. From partitions of this matrix we can compute $\mathbb{D}(\lambda_E, \lambda_B)(2 \times 2)$ by using Equation 5.69.

To illustrate this, let us choose specific values and use a computer to perform the required calculations for the parameters k, $N_0/2$, P, and T. Let us set

$$k = 1, \tag{5.83a}$$

$$N_0/2 = 1, \tag{5.83b}$$

$$P = 1, \tag{5.83c}$$

$$T = 2. \tag{5.83d}$$

Since this spectrum is generated by a first-order system, we want to plot the largest two loci, i.e., for a given value of λ_E, the points on these loci are the largest two values of λ_B solving Equation 5.71. We point out that other solutions with smaller values of λ_B exist. These loci are illustrated in Figure 5.1.

The reason for considering the asymptotic solutions should now be apparent. We have sketched their loci as given by Equation 5.62 with dotted lines. We see that the loci asymptotically approach those of the asymptotic solutions. Therefore, for large values of λ_E a convenient place to start searching for the loci is near those specified by Equation 5.62. We should point out that the loci of the asymptotic solutions do not cause the determinant specified by Equation 5.71 to vanish. This is because in determining their existence we used the equations directly derived from the minimum principle rather than the reduced set of equations that were used to derive Equation 5.71.

Now, if we take the solutions specified by these loci and determine the bandwidth and performance of the corresponding signal according to steps (b) through (e), we produce a second set of loci in a d_g^2-B^2 plane. These are illustrated in Figure 5.2. In addition we have indicated the bandwidth and performance of the asymptotic solutions by a large dot near the identifying number of the loci that approaches it. (We can

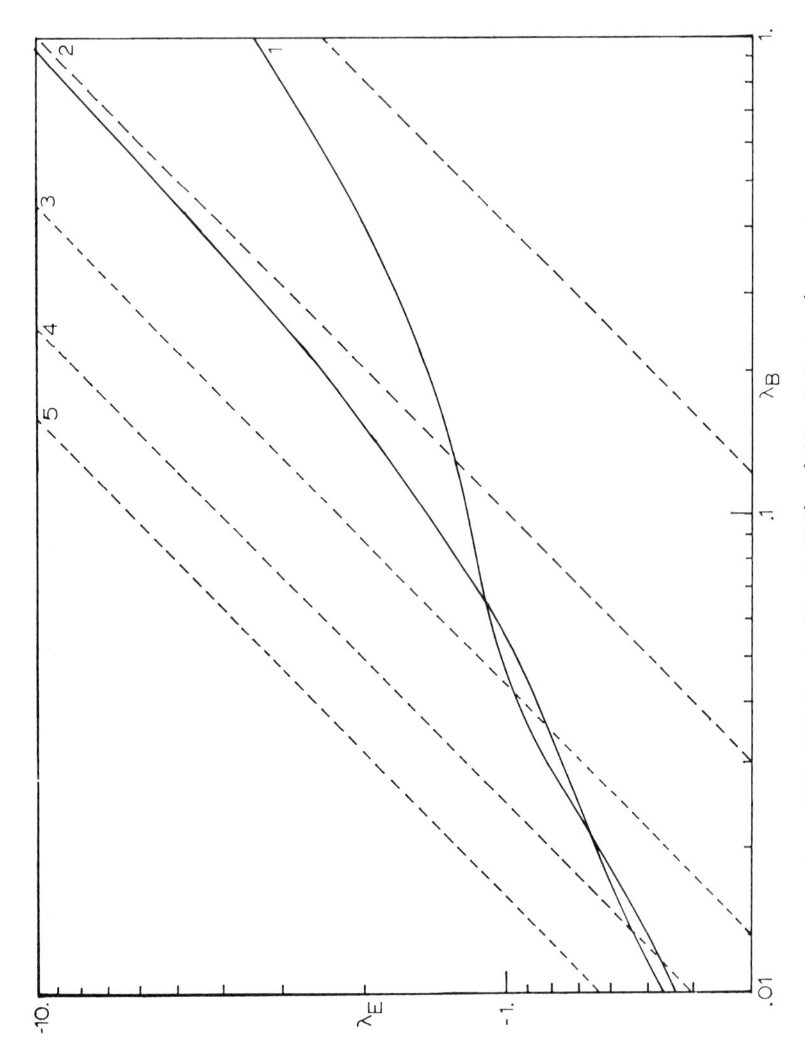

Figure 5.1 Solution loci for $\det[\mathbf{D}(\lambda_E, \lambda_B)] = 0$, first-order spectrum.

indicate it by a single dot because the entire loci for the asymptotic solution corresponds to just one signal.)

In the λ_E-λ_B plane of Figure 5.2, as we move from left to right on the solid lines, which are those that cause the determinant of Eq. 5.71 to vanish, we generate the loci in the d_g^2-B^2 plane with the solid lines in the direction indicated by the arrowheads. We see that they evolve from the dots corresponding to the asymptotic solutions. (If one plots the loci corresponding to suboptimal solutions, their behavior are not nearly as well behaved. One and two are well behaved, while those corresponding to the remaining numbers have a rather erratic behavior.

As indicated by step (f), the other aspect of the optimization that needs to be considered is the crossings of the loci that occur in the λ_E-λ_B plane. At these crossings, there are two signals with the same values of λ_E and λ_B which satisfy the necessary conditions. If we find the degradation and bandwidth of all possible linear combinations of these signals (normalized to unit energy), we produce the loci in the d_g^2-B^2 plane indicated by the dotted lines. (There is no relation between the dotted lines in each plane.) Once we have generated these loci over the bandwidth constraint region of interest, we can find the optimal signal corresponding for any particular bandwidth constraint value. We merely select the signal that produces the absolute minimum degradation, or the lowest point in the d_g^2-B^2 plane for the specified value of B.

Let us illustrate the actual form of some of the optimal signals for two different values of the bandwidth constraint, B. In the first illustration we have B^2 equal to 9. If we examine the d_g^2-B^2 of Figure 5.3, we see that the optimal signal is a linear combination of the two solutions that are produced by the first crossing of loci one and two in the λ_E-λ_B plane. The optimal signal for this constraint is illustrated in Figure 5.3. We see that we can achieve a degradation of 23.6 per cent compared to the white-noise-only performance.

In addition, we have drawn $g(t)$, the correlation signal for the optimal receiver. The signal(s) exhibits no particular symmetry for this constraint value of B equal to 3 since it is composed principally of the signals similar in form to $\sin(\pi t/T)$ and $\sin(2\pi t/T)$. We should note that because of the possibility of two different sign reversals there are actually four signals that are optimal. All of these, however, basically have the same waveshape.

In Figure 5.4 we show the optimal signal $s(t)$ and its correlating signal $g(t)$ when we allow twice as much bandwidth, i.e., $B^2 = 36$. In this case, the optimal signal does not correspond to a crossing in the λ_E-λ_B plane, as it did previously. The bandwidth is sufficiently large

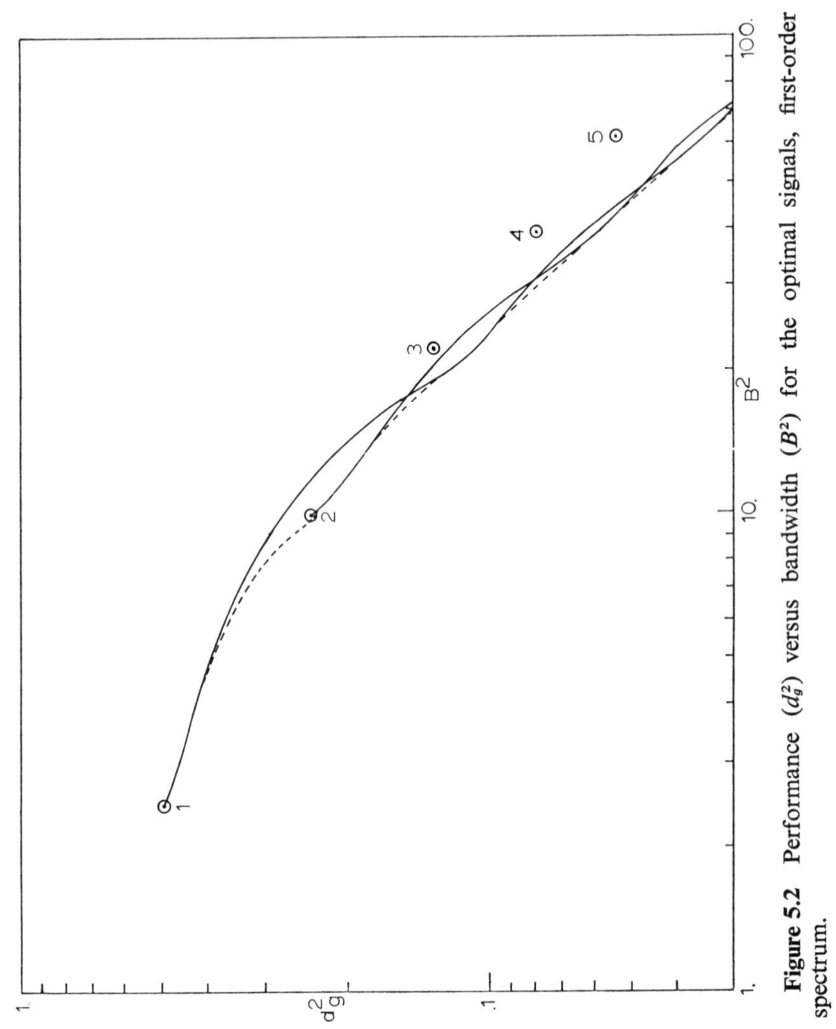

Figure 5.2 Performance (d_g^2) versus bandwidth (B^2) for the optimal signals, first-order spectrum.

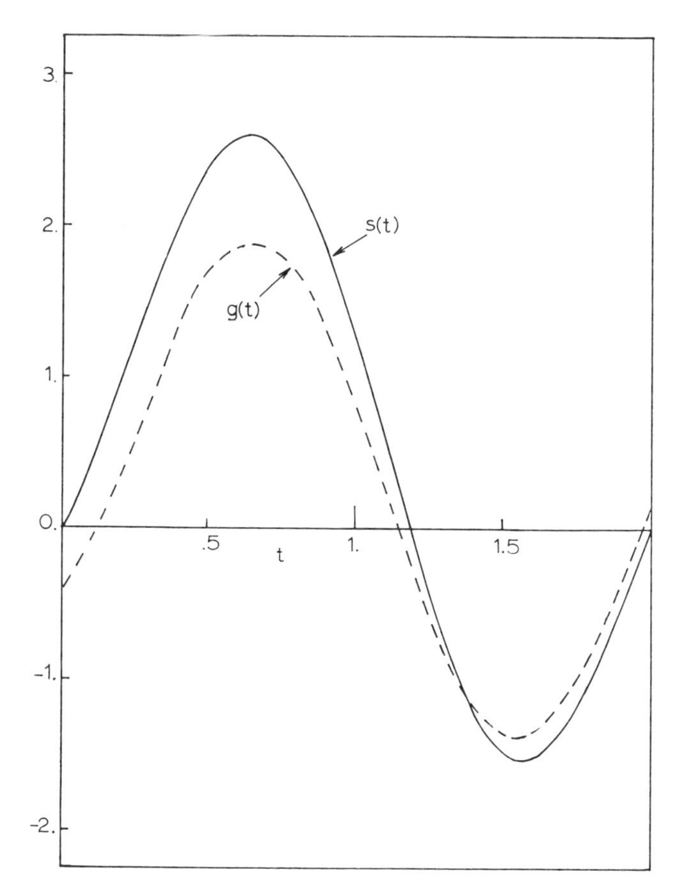

Figure 5.3 Optimal $s(t)$ and $g(t)$ for a first-order spectrum $B^2 = 9$, $d_g^2 =$ 0.265.

enough so that we can attain a performance only 6.4 per cent below that of only the white noise being present. The signal does display some symmetry in that $s(t) = -s(T - t)$; and the correlating signal is almost identical to $s(t)$ indicating that we (nearly) have a white-noise-type receiver.

We conclude this example by discussing the improvement over a more conventional signal which is attainable by transmitting an optimal signal. We propose four transmission schemes as shown in Table 5.1. In the first two methods we transmit a pulsed sine wave with unit energy, i.e., $\sqrt{2/T}\sin(n\pi t/T)$. The bandwidth that this signal consumes is $n\pi/T$.

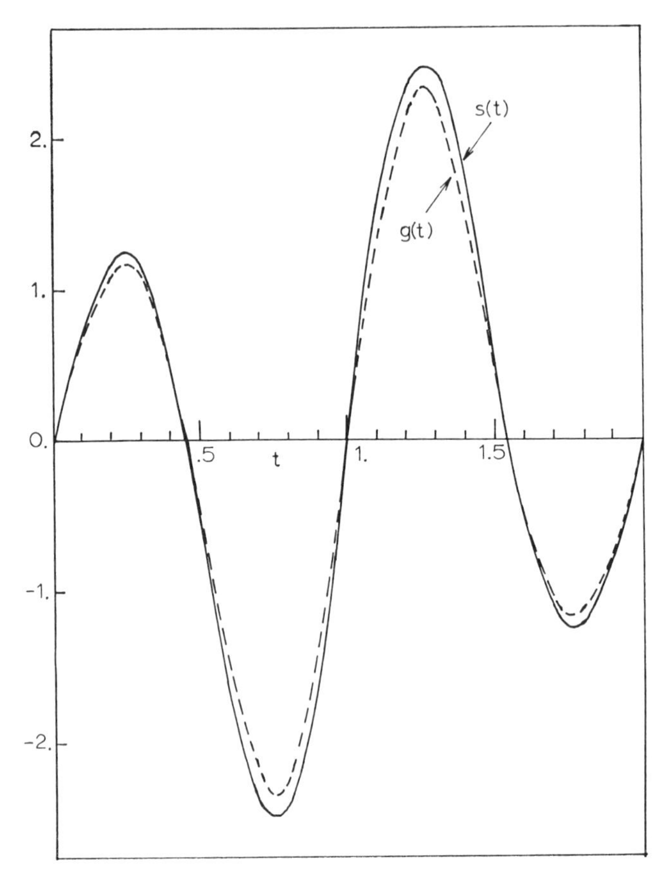

Figure 5.4 Optimal $s(t)$ and $g(t)$ for a first-order spectrum $B^2 = 36$, $d_g^2 = 0.0643$.

For the receiver in the first scheme, we use a conventional matched filter, while in the second we use an optimal correlating signal $g_n(t)$ as discussed in Chapter 4. In the second two methods we transmit optimally designed signals that consume the same bandwidth as the pulsed sine waves. In the third scheme, we optimally design the signals when the receiver is constrained to be a matched filter. This problem has been analyzed by Post.[48] In the fourth scheme we optimally design the signals, as we have discussed. We make this comparison for the first five values of n and two levels of colored noise. The first level is for $P = 1$, while the second is for $P = 10$, which increases the colored noise component power by a factor of 10, such that it dominates the spectrum.

Table 5.1 Comparison of Performance for Optimally Designed Signals

n	$B\left(\frac{\text{rad}}{\text{sec}}\right)$	Pulsed Sine Wave Matched Filter RCVR d^2	d_g^2	Pulsed Sine Wave Optimum RCVR d^2	d_g^2	Optimal Signal for Matched Filter RCVR[a] d^2	d_g^2	Optimal Signal for Optimum RCVR d^2	d_g^2
$P = 1$									
1	1.57	0.490	0.510	0.510	0.490	0.490	0.510	0.510	0.490
2	3.14	0.753	0.247	0.761	0.239	0.753	0.247	0.765	0.235
3	4.71	0.848	0.152	0.870	0.130	0.891	0.109	0.895	0.105
4	6.28	0.917	0.083	0.922	0.078	0.942	0.058	0.943	0.057
5	7.85	0.937	0.063	0.947	0.053	0.964	0.036	0.964	0.037
$P = 10$									
1	1.57	0.088	0.912	0.116	0.884	0.088	0.912	0.116	0.884
2	3.11	0.234	0.766	0.278	0.722	0.234	0.766	0.297	0.703
3	4.72	0.359	0.641	0.452	0.548	0.450	0.550	0.488	0.512
4	6.28	0.525	0.475	0.592	0.408	0.619	0.381	0.637	0.363
5	7.85	0.598	0.402	0.693	0.307	0.734	0.266	0.748	0.252

[a] From Post.[48]

We can see that, for $P = 1$, the increase in performance is hardly outstanding. For $P = 10$, we can realize about a 5–10 per cent improvement with the optimum system as compared to a pulsed sine wave with an optimal receiver and 15–20 per cent improvement over a pulsed sine wave with a matched filter. One might possibly guess this, since the first-order noise process does not have much structure to exploit by designing the signals correctly. For example, it is the most difficult of the Butterworth processes to filter.

Example 5.2 Signal Design with a Second-Order Spectrum

In this example we consider the signal design problem when the colored component of the observation noise has a second-order spectrum. We do this for two reasons. First, we want to develop a better understanding of the problems involved in implementing our algorithm. Second, we want to demonstrate that our optimization procedure can produce more significant improvements when working with a process that has more "structure" to it.

We assume that the colored component of the observation noise is generated by Equation 2.28. In addition, we set the white-noise spectral level to unity and assume that the interval length is two, i.e., $T_0 = 0$,

$T_f = T = 2$. Hence, the colored component of the observation noise dominates for all frequencies less than approximately 0.8 Hz.

This particular colored noise spectrum is particularly interesting because of its shape. Suppose we constrain the bandwidth to be less than 3, or $B^2 = 9$. This corresponds to allowing frequencies that are lower than that frequency where the spectrum has its peak. We certainly do not want to put all the signal energy in frequencies near the peak. Yet, if the bandwidth is available we should be able to use it to our advantage. Although one may conjecture from either our previous example, or from one's engineering judgment, that a linear combination of pulsed sine waves should be close to optimum, it is not very apparent that this is so.

Let us proceed with the steps in our algorithm. The first three loci in the λ_E-λ_B plane are illustrated and numbered in Figure 5.5. Again, we observe the asymptotic behavior for large negative values of λ_E. In two respects the loci exhibit a somewhat more complicated behavior than those for the first-order spectrum. First, they do not cross simply pairwise. Apparently, they cross with the adjacent locus; i.e., from plotting more loci we find that the second crosses the first and third, the third crosses the second and fourth, etc. Second, in some regions they are extremely close together.

In Figure 5.6 we have plotted the corresponding loci in the d_g^2-B^2 plane for the largest two loci in the λ_E-λ_B plane. Their behavior is very similar to the first-order case. As indicated by the dotted lines, there are regions where we must consider the linear combinations of signals corresponding to a loci crossing in the λ_E-λ_B plane.

Let us now examine the shape of an optimal signal when we constrain the bandwidth such that the colored noise is dominant over the allowed frequency range. If we chose $B^2 = 13$, the optimum signal $s(t)$ and its correlating signal $g(t)$ are as illustrated in Figure 5.7. The signal is principally composed of functions of the form $\sin(\pi t)$ and $\sin(3\pi t)$. The degradation realized by this signal is 0.53, which is approximately 15 per cent below what one might expect using a nonoptimum approach of selecting the pulsed sine wave with the best performance that still satisfies the bandwidth constraint.

In Table 5.2 we have summarized the performance in a way similar to that done in the first example. We see that in some cases we can improve our performance by approximately 20 per cent. However, if we were to compare performance to a linear combination of pulsed sine waves, e.g., the first and third, the improvement would not be as significant.

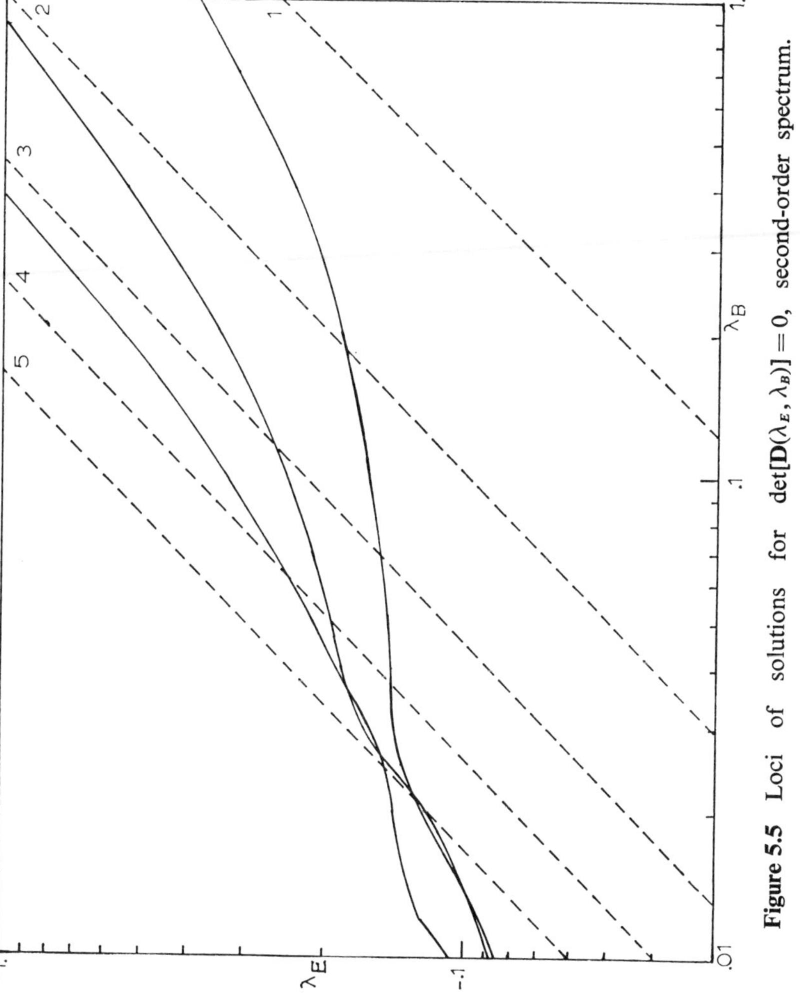

Figure 5.5 Loci of solutions for $\det[\mathbf{D}(\lambda_E, \lambda_B)] = 0$, second-order spectrum.

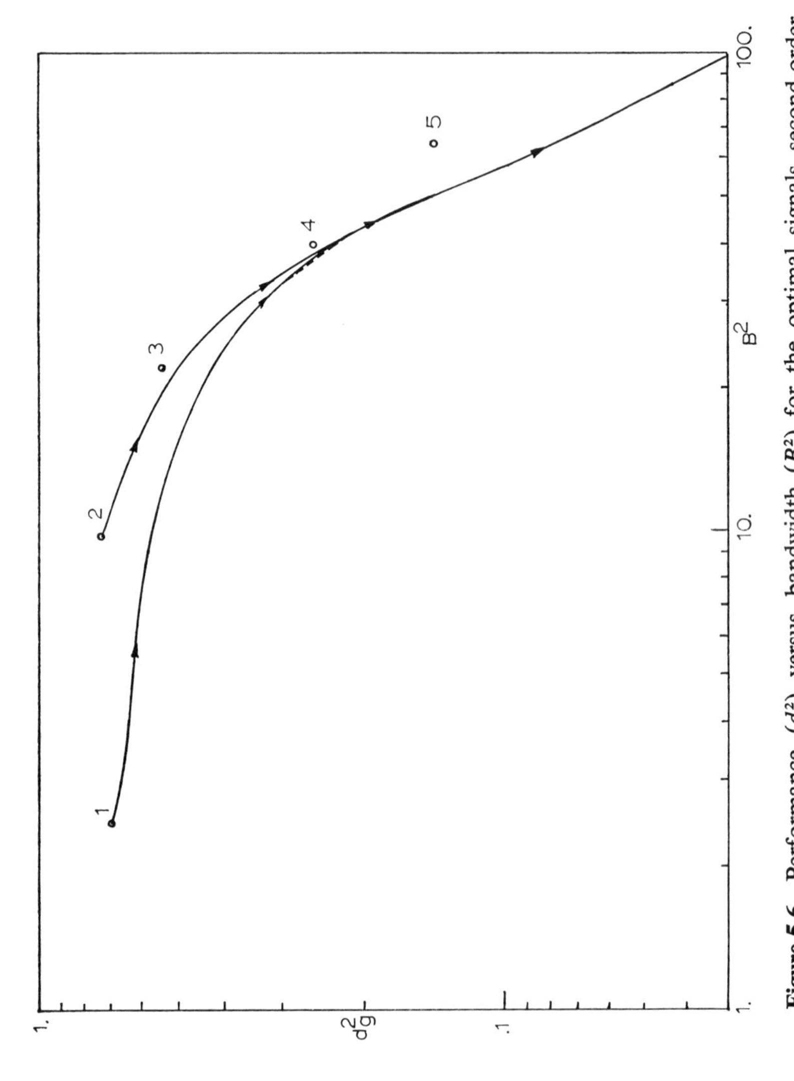

Figure 5.6 Performance (d_g^2) versus bandwidth (B^2) for the optimal signals, second-order spectrum.

Table 5.2 Comparison of Performance for Optimally Designed Signals

		Pulsed Sine Wave Matched Filter RCVR		Pulsed Sine Wave Optimum RCVR		Optimum Signal Optimum RCVR	
n	$B\left(\frac{\text{rad}}{\text{sec}}\right)$	d^2	d_g^2	d^2	d_g^2	d^2	d_g^2
1	0.57	0.31	0.69	0.32	0.68	0.32	0.68
2	3.14	0.26	0.74	0.27	0.73	0.43	0.57
3	4.72	0.40	0.60	0.41	0.59	0.58	0.42
4	6.28	0.70	0.30	0.75	0.25	0.77	0.23
5	7.85	0.50	0.20	0.86	0.14	0.92	0.08

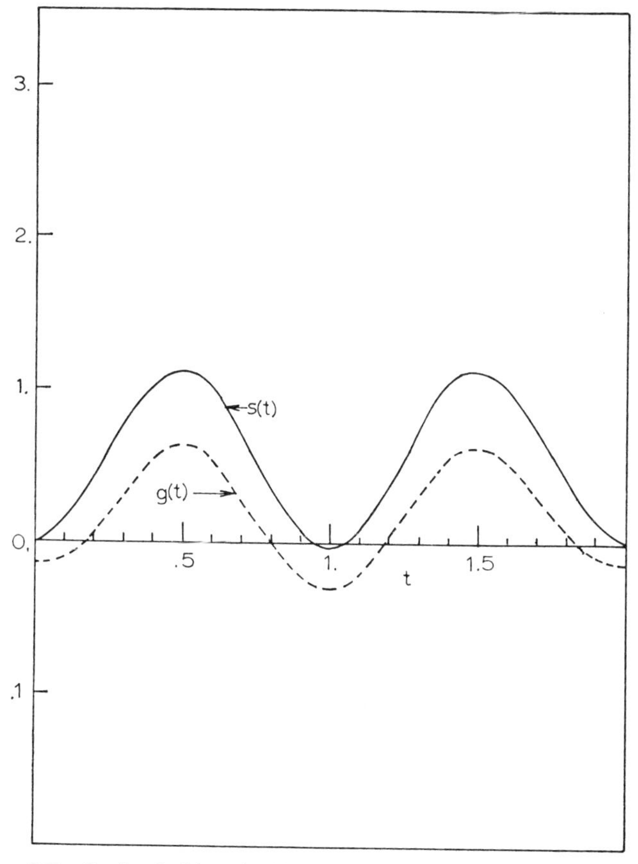

Figure 5.7 Optimal $s(t)$ and $g(t)$ for a second-order spectrum $B^2 = 13$.

Signal Design with a Hard Bandwidth Constraint

Before concluding this section we derive the differential equations that specify the necessary condition for optimality when we constrain the absolute value of the derivative of the signal, i.e., we require

$$|v(t)| = \left|\frac{ds(t)}{dt}\right| \leqq E^{1/2}B, \qquad T_0 \leqq t \leqq T_f. \tag{5.84}$$

As before, we constrain the signal energy by

$$\int_{T_0}^{T_f} s^2(\tau)\, d\tau \leqq E. \tag{5.85}$$

We can formulate the problem in a manner very similar to that used previously. The resulting Hamiltonian is

$$\begin{aligned} &H(\boldsymbol{\xi}, \boldsymbol{\eta}, s, x_E, \mathbf{p}_{\xi}, \mathbf{p}_{\eta}, p_s, \lambda_E, v) \\ &= p_0 \frac{2}{N_0} s(t)\mathbf{C}(t)\boldsymbol{\xi}(t) + \mathbf{p}_{\xi}^T(t)(\mathbf{F}(t)\boldsymbol{\xi}(t) + \mathbf{G}(t)\mathbf{Q}\mathbf{G}^T(t)\boldsymbol{\eta}(t)) \\ &\quad + \mathbf{p}_{\eta}^T(t)\left(\mathbf{C}^T(t)\frac{2}{N_0}\mathbf{C}(t)\boldsymbol{\xi}(t) - \mathbf{F}^T(t)\boldsymbol{\eta}(t) - \mathbf{C}^T(t)\frac{2}{N_0}s(t)\right) \\ &\quad + p_s(t)v(t) + \frac{\lambda_E(t)}{2}s^2(t). \end{aligned} \tag{5.86}$$

Taking the required derivatives, we find

$$\frac{\partial H}{\partial \boldsymbol{\xi}} = -\dot{\mathbf{p}}_{\xi}(t) = p_0\, \mathbf{C}^T(t)\frac{2}{N_0}s(t) + \mathbf{F}^T(t)\mathbf{p}_{\xi}(t) + \mathbf{C}^T(t)\frac{2}{N_0}\mathbf{C}(t)\mathbf{p}_{\eta}(t), \tag{5.87}$$

$$\frac{\partial H}{\partial \boldsymbol{\eta}} = -\dot{\mathbf{p}}_{\eta}(t) = \mathbf{G}(t)\mathbf{Q}\mathbf{G}^T(t)\mathbf{p}_{\xi}(t) - \mathbf{F}(t)\mathbf{p}_{\eta}(t), \tag{5.88}$$

$$\frac{\partial H}{\partial s} = -\dot{p}_s(t) = p_0 \frac{2}{N_0}\mathbf{C}(t)\boldsymbol{\xi}(t) - \frac{2}{N_0}\mathbf{C}(t)\mathbf{p}_{\eta}(t) + \lambda_E(t)s(t), \tag{5.89}$$

$$\frac{\partial H}{\partial E} = -\dot{\lambda}_E(t) = 0. \tag{5.90}$$

The transversatility conditions imply

$$\mathbf{P}_0\, \mathbf{p}_{\xi}(T_0) = -\mathbf{p}_{\eta}(T_0), \tag{5.91}$$

$$\mathbf{p}_{\xi}(T_f) = 0. \tag{5.92}$$

The $p_s(T_f)$ and $p_s(T_0)$ are free. We again have that λ_E is a constant. Furthermore, we note that Equations 5.87, 5.88, 5.91, and 5.92 are identical with Equations 5.36a, 5.36b, 5.36c, and 5.36d, respectively. Consequently, we can again show by using the results of Section 5.3 that

$$\boldsymbol{\xi}(t) = -\mathbf{p}_{\boldsymbol{\eta}}(t), \tag{5.93a}$$

$$\boldsymbol{\eta}(t) = \mathbf{p}_{\boldsymbol{\xi}}(t). \tag{5.93b}$$

Therefore Equation 5.89 becomes (assuming p_0 equals unity)

$$\dot{p}_s(t) = -\frac{4}{N_0}\mathbf{C}(t)\boldsymbol{\xi}(t) - \lambda_E s(t). \tag{5.94}$$

The major difference between the application of the minimum principle to this problem and the one in the text comes in the minimization of the Hamiltonian as a function of the control $v(t)$. If $p_s(t)$ is nonzero, this minimization implies

$$v(t) = -E^{1/2}B \operatorname{sgn}(p_s(t)). \tag{5.95}$$

This implies that the optimal signal has a constant linear slope of $\pm E^{1/2}B$ when $p_s(t)$ is nonzero.

The differential equations, now nonlinear, that specify the necessary conditions are, for $T_0 \leqq t \leqq T_f$,

$$\frac{d\boldsymbol{\xi}(t)}{dt} = \mathbf{F}(t)\boldsymbol{\xi}(t) + \mathbf{G}(t)\mathbf{Q}\mathbf{G}^T(t)\boldsymbol{\eta}(t), \tag{5.96}$$

$$\frac{d\boldsymbol{\eta}(t)}{dt} = \mathbf{C}^T(t)\frac{2}{N_0}\mathbf{C}(t)\boldsymbol{\xi}(t) - \mathbf{F}^T(t)\boldsymbol{\eta}(t) - \mathbf{C}^T(t)\frac{2}{N_0}s(t), \tag{5.97}$$

$$\frac{ds(t)}{dt} = v(t) = -E^{1/2}B \operatorname{sgn}(p_s(t)), \tag{5.98}$$

$$\frac{dp_s(t)}{dt} = -\frac{4}{N_0}\mathbf{C}(t)\boldsymbol{\xi}(t) - \lambda_E s(t). \tag{5.99}$$

The boundary conditions are

$$\boldsymbol{\xi}(T_0) = \mathbf{P}_0\boldsymbol{\eta}(T_0), \tag{5.100}$$

$$\boldsymbol{\eta}(T_f) = \mathbf{0}, \tag{5.101}$$

$$s(T_0) = s(T_f) = 0. \tag{5.102}$$

By multiplying Equation 5.99 by $s(t)$ and then integrating by parts, we can demonstrate that

$$E^{1/2}B\int_{T_0}^{T_f}|p_s(t)|\,dt + 2d_g^2 + \lambda_E E = 0. \tag{5.103}$$

This implies that λ_E is negative and also provides a convenient way of finding d_g^2.

If $p_s(t)$ is zero over any region, we cannot determine $v(t)$ by Equation 5.95. In such a region

$$p_s(t) = 0 \tag{5.104}$$

implies

$$s(t) = -\frac{4}{\lambda_E N_0}\,\mathrm{C}(t)\xi(t). \tag{5.105}$$

If we combine this equation with Equation 5.96, we find that the two equations have the same form as those that specify the eigenfunctions associated with the colored noise.

This leads us to the following conjecture: The optimal signal consists of regions of a constant slope of $E^{1/2}B$ and regions where the signal has the same functional form as the eigenfunction of the colored noise. At this point we probably need to resort to numerical procedures for actually determining a solution. In Reference 2 a survey of several techniques for implementing this are discussed.

5.4 Optimal Signal Design for Doppler Spread Environments

In many environments the noise is multiplicative, such that we have for the noise process

$$y(t) = b(t)s(t). \tag{5.106}$$

We therefore have the detection problem*

$$\begin{aligned} H_1:\quad r(t) &= s(t) + b(t)s(t) + w(t), \\ H_0:\quad r(t) &= \phantom{s(t) + {}} b(t)s(t) + w(t). \end{aligned} \tag{5.107}$$

* To be more realistic, this problem should be formulated at bandpass using the complex notation of Appendix B with a Rayleigh fading for the signal such that we have

$$\begin{aligned} H_1\quad \tilde{r}(t) &= \tilde{b}\tilde{s}(t) + \tilde{b}(t)\tilde{s}(t) + \tilde{w}(t), \\ H_0\quad \tilde{r}(t) &= \phantom{\tilde{b}\tilde{s}(t) + {}} \tilde{b}(t)\tilde{s}(t) + \tilde{w}(t), \end{aligned}$$

where b is a complex random variable. As indicated in Chapter 4, the figure of merit is still d^2 and the issues in the optimization are essentially the same.

This is often termed a Doppler spread, or multiplicative, fading environment.

Before concluding this section, we simply point out how our results in Section 5.2 can be used to specify the necessary conditions for the existence of an optimal signal when we use the mean square constraints for the energy and bandwidth of the signal.

To apply the results derived there, we model $b(t)$ to be Gaussian random process as described in Chapter 2. If $x(t)$ is the state vector for the system we have

$$b(t) = \mathbf{C}_0(t)\mathbf{x}(t). \tag{5.108}$$

Consequently, the noise process $y(t)$ becomes

$$y(t) = s(t)\mathbf{C}_0(t)\mathbf{x}(t). \tag{5.109}$$

Therefore, using our earlier notation, we have for the observation matrix $C(t : s(t))$ describing the generation of $y(t)$

$$C(t : s(t)) = s(t)\mathbf{C}_0(t). \tag{5.110}$$

Consequently, we obtain

$$\frac{\partial \mathbf{C}(t : s(t))}{\partial s(t)} = \mathbf{C}_0(t), \tag{5.111}$$

and the necessary conditions for an optimal signal as expressed by Equations 5.48 and 5.49 become

$$\frac{d\boldsymbol{\xi}(t)}{dt} = \mathbf{F}(t)\boldsymbol{\xi}(t) + \mathbf{G}(t)\mathbf{Q}\mathbf{G}^T(t)\boldsymbol{\eta}(t), \tag{5.112a}$$

$$\frac{d\boldsymbol{\eta}(t)}{dt} = s^2(t)\mathbf{C}_0(t)\frac{2}{N_0}\mathbf{C}_0^T(t)\boldsymbol{\xi}(t) - \mathbf{F}^T(t)\boldsymbol{\eta}(t) - \frac{2}{N_0}\mathbf{C}_0(t)s^2(t), \tag{5.112b}$$

$$\frac{ds(t)}{dt} = (-1/\lambda_B)p_s(t), \tag{5.112c}$$

$$\frac{dp_s(t)}{dt} = -\frac{4}{N_0}s(t)\mathbf{C}_0(t)\xi(t) + \frac{4}{N_0}s(t)(\mathbf{C}_0(t)\boldsymbol{\xi}(t))^2 - \lambda_E s(t),$$

$$T_0 \leqq t \leqq T_f. \tag{5.112d}$$

$$\boldsymbol{\xi}(T_0) = \mathbf{P}_0\boldsymbol{\eta}(T_0), \tag{5.113a}$$

$$\boldsymbol{\eta}(T_f) = 0, \tag{5.113b}$$

$$s(T_0) = s(T_f) = 0. \tag{5.113c}$$

At this point we have reduced it to a set of equations which needs to be solved by numerical methods using a computer. Again several methods have been developed in the literature, and a convenient reference is Reference 2. There are several limiting situations that can be discussed depending upon the bandwidth of the fading, the constraint bandwidth, and the duration of the signal. In many cases the problem can be approximated by a simpler one involving only random variables. See Van Trees for a more complete discussion.[68] Probably the most interesting situation using the state variable formulation is when the constraint and fading bandwidth are approximately equal.

5.5 Summary and Discussion

We have presented a state variable method for designing optimal signals for detection in colored noise when there are energy and bandwidth constraints. The performance measure was given by d_g^2, which specified the loss of receiver performance due to the colored noise being present. We used the differential equations and their associated boundary condition that specified the optimal receiver and performance measure as if they described a dynamic system. We then applied Pontryagin's minimal principle to derive the necessary conditions that the optimal signal must satisfy. These conditions specified a characteristic value problem. We can determine the optimal signal by solving this characteristic value problem.

Although these conditions were valid for the signal imbedded in the observation matrix in a rather arbitrary no-memory manner, we principally confined our attention to the situation when this matrix was not a function of $s(t)$, with the result that the noise was independent of the signal. Here we were able to take the necessary conditions and develop an algorithm for actually finding the optimal signals. By using this algorithm we were able to analyze two examples of colored noise spectra. In addition to finding the optimal signal and its associated performance, the algorithm displayed several interesting features.

One may argue that we did not need to use the minimum principle to solve this problem since we could proceed directly to the $2n + 2$ differential equations and boundary conditions via the usual calculus of variations and the estimator-subtractor realization of the receiver. The advantage of this formulation became apparent when we changed the type of constraints upon the signal. We derived the differential equations that specify the optimal signal when we imposed an energy constraint and a hard (bandwidth) constraint, $|ds(t)|dt| \leqq B$. This problem is readily approached using a minimum principle, whereas the calculus of variations would require much more effort.

There are several other important issues.

1. In the signal-independent noise problem we were able to find a relation between λ_E, λ_B, E, B^2, and d_g^2. This was very useful in implementing our algorithm. A more general statement of this result would be valuable when the noise is signal dependent.

2. The equivalence of $\boldsymbol{\xi}(t)$ and $-\mathbf{p}_{\eta}(t)$, and $\boldsymbol{\eta}(t)$ and $\mathbf{p}_{\xi}(t)$ is a very general result. Because of this generality it seems that there should be a means of formulating these problems with a stochastic minimum principle which would allow us to obtain the final differential equations directly.[41] However, this means has not been apparent so far.

3. Finally, the results in this section can be extended to bandpass signal design by using the material in Appendix B. The possibility of a noise spectrum that is not symmetrical about the carrier introduces some interesting questions on how to best take advantage of this situation.

4. Although our results are valid for an arbitrary signal duration, it becomes difficult to implement them when the time bandwidth product allowed for the signal becomes large. For stationary processes, the problem essentially degenerates into one that can be quickly solved by spectral analysis when the "$2WT$" product becomes large, much as the eigenvalues approach a stationary distribution. When the noise is nonstationary or signal dependent, this issue is not so apparent, basically because of the nonstationarity.

Our basic approach to signal design is to couple our Fredholm integral equation theory to optimal control theory. This theory is relevant to a large number of problems and our approach can be used in a large number of them. Signal design has been approached in several ways in the recent literature. The work of Schweppe and Athans is probably most similar to ours, being strongly coupled to state variable concepts.[5] They use many of the same assumptions, but couple their approach to the realizable filter theory. Holtzman has considered the optimal design of signals for transmission through a channel filter so as to maximize the output signal to noise ratio using state variables.[32] Tufts and Shnidman have also approached the problem when hard constraints are imposed.[65] Titlebaum and Thompson have approached the problem of optimal receiver design for clutter suppression using state variables.[63] Schweppe and Gray have also considered optimum waveform design for radar tracking.[53]

Classically, one has the smallest eigenvalue approach as discussed by Middleton and later by Van Trees for the signal-independent noise channel.* For signal-dependent noise, or clutter environments, Van Trees has discussed a very intuitive technique involving the scattering function of the noise and the ambiguity function of the signal.[66] Other recent work has been done by Delong and Hofstetter,[23] by Rummler,[57] by Spafford,[60] by Kincaid,[38, 39] and by Balakrishnan.[12] To date, the results in the literature do not seem conclusive, especially with regard to the global aspects of the optimization problem.

Kennedy has found bounds on the performance of any signal when one communicates over a spread channel.[37] This is particularly important since it sets a limit, or benchmark, on the potential payoff using optimally designed signals. In some representative problems, some very simple signals perform quite close to the bound. Finally we have considered only binary situations. The M-Ary problem for white-noise channels has been approached by Weber, using geometric interpretations.[71]

* Ref. 67, pp. 302.

6 Linear Smoothing and Filtering with Delay

In this chapter we use the results derived in Chapter 4 for solving inhomogeneous Fredholm integral equations to develop a unified approach to linear smoothing and filtering with delay. In the smoothing problem we receive a signal over a fixed time interval $[T_0, T_f]$. We then want to find $\hat{\mathbf{x}}(t)$, $T_0 \leqq t \leqq T_f$, the estimate of the state vector that generated this signal over the same fixed interval. This is illustrated in the left side of Figure 6.1. In a filtering problem we receive a signal continuously, i.e., the end point time of the interval, T_f, is constantly increasing. We then want to produce an estimate that evolves in time as a function of this end point. For the realizable filter we want to find $\hat{\mathbf{x}}(T_f)$ versus T_f, the estimate of the state vector right at the end point time, T_f. For the filter with delay, or lag, we are allowed a delay before we make our estimate. We want to find $\hat{\mathbf{x}}(T_f - \Delta)$ versus T_f, the estimate of the state vector Δ units prior to the end point of the interval. This is illustrated in the right side of Figure 6.1.

Our approach to these problems is straightforward. We start with the Wiener-Hopf equation that specifies the impulse response for the optimal linear estimator.* We then show how we can find a set of differential equations that specifies the optimal estimate implicitly as part of its solution. From these equations we can derive matrix differential equations that determine the covariance of error.

Our approach to the problem of filtering with delay is also straight-

* Ref. 67, pp. 472.

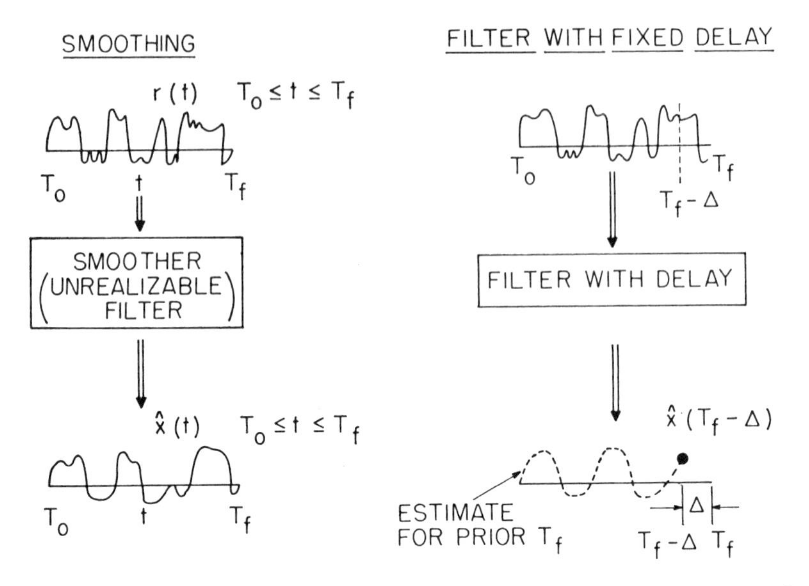

Figure 6.1 Comparison of the smoothing (interval estimation) and filtering with fixed-delay estimation problems.

forward. In the solution to the optimal smoother, we simply allow the end point time of the observation interval to be a variable. We then derive a set of differential equations that is a function of this variable end point time rather than the time within a fixed observation interval as for the smoother. The performance is also derived in an analogous manner.

Several comments are in order before proceeding. First, the most important point to be made concerning our methods is the approach. The entire theory is developed concisely and directly starting from a few basic results.

Second, we employ a structured approach to this topic. We require the estimator structure to be linear, regardless of the statistics of the processes involved. Most of the existing approaches to this problem are unstructured.[15,49] It is assumed that the processes involved are Gaussian, and then the estimator structure is derived. It is well known, however, that both approaches lead to the same estimator.

Third, we assume that the reader is familiar with the well-known results for realizable filtering by using state variable techniques, i.e., the Kalman-Bucy filter.[35]

6.1 The Optimal Linear Smoother

In this section, we derive a state variable realization of the optimum linear smoother, the state variable equivalent of the unrealizable filter. Our derivation follows Reference 7. First, we establish our model.

Let us assume that we generate a random process $\mathbf{y}(t)$ by the methods described in Chapter 2. Let us also assume that we observe this process in the presence of an additive white noise $\mathbf{w}(t)$ over the interval $T_0 \leqq t \leqq T_f$. That is, we receive the signal

$$\mathbf{r}(t) = \mathbf{y}(t) + \mathbf{w}(t) = \mathbf{C}(t)\mathbf{x}(t) + \mathbf{w}(t), \qquad T_0 \leqq t \leqq T_f, \tag{6.1}$$

where $\mathbf{x}(t)$ is the state vector of the system that generates $\mathbf{y}(t)$; $\mathbf{w}(t)$ is an additive white observation noise that has a covariance $\mathbf{R}(t)\delta(t-\tau)$ as given by Equation 4.1.

In the optimal smoothing problem we want to find a state variable description of the linear system that minimizes the mean-square error in estimating each component of the state vector $\mathbf{x}(t)$. This description we will find consists of two first-order vector differential equations having a two-point boundary restriction.

First, we define the matrix impulse response of the optimal linear smoother to be $\mathbf{h}_0(t, \tau)$, or,*†

$$\hat{\mathbf{x}}(t) = \int_{T_0}^{T_f} \mathbf{h}_0(t, \tau)\mathbf{r}(\tau)\, d\tau, \qquad T_0 \leqq t \leqq T_f. \tag{6.2}$$

This operator produces an estimate of the state vector $\mathbf{x}(t)$ at time t by observing $\mathbf{r}(\tau)$ over the entire interval $[T_0, T_f]$.

It is well known and can easily be shown by classical methods that this impulse response satisfies the following Wiener-Hopf integral equation:

$$\mathbf{K}_{\mathbf{dr}}(t, \tau) = \int_{T_0}^{T_f} \mathbf{h}_0(t, v)\mathbf{K}_{\mathbf{r}}(v, \tau)\, dv, \qquad T_0 \leqq t, \tau \leqq T_f, \tag{6.3}$$

where $\mathbf{K}_{\mathbf{dr}}(t, \tau)$ is the cross covariance of the desired signal and the received signal and $\mathbf{K}_{\mathbf{r}}(t, \tau)$ is the covariance of the received signal. For our application, the desired signal is the state vector $\mathbf{x}(t)$. Therefore, we have

$$\mathbf{K}_{\mathbf{x}}(t, \tau)\mathbf{C}^T(\tau) = \int_{T_0}^{T_f} \mathbf{h}_0(t, v)\mathbf{K}_{\mathbf{r}}(v, \tau)\, dv, \qquad T_0 \leqq t, \tau \leqq T_f, \tag{6.4}$$

* Although $\mathbf{h}_0(t, \tau)$ is a matrix and should be denoted by a capital letter in our notation convention, we defer to the conventional notation.

† We have assumed zero means for $\mathbf{x}(T_0)$, $\mathbf{u}(t)$ and $\mathbf{w}(t)$. If the means were nonzero, we would need to add a bias term to Equation 6.2.

with

$$\mathbf{K}_\mathbf{r}(t, \tau) = \mathbf{K}_\mathbf{y}(t, \tau) + \mathbf{R}(t)\delta(t - \tau). \tag{6.5}$$

The first step in our derivation is to solve this equation for $\mathbf{h}_0(t, \tau)$. In order to do this, we need to introduce the inverse kernel $\mathbf{Q}_\mathbf{r}(t, \tau)$ of $\mathbf{K}_\mathbf{r}(t, \tau)$, the covariance of the received signal.

Let us now introduce some material that we need from Chapter 4. From Equation 4.3 we can write the inhomogeneous Fredholm integral equation as

$$\int_{T_0}^{T_f} \mathbf{K}_\mathbf{r}(t, \tau)\mathbf{g}(\tau) = \mathbf{s}(t), \qquad T_0 \leqq t \leqq T_f, \tag{6.6}$$

As we discussed in Chapter 4, we can consider that this integral equation specifies a linear operator upon $\mathbf{s}(t)$, with the solution $\mathbf{g}(t)$ being the result of this linear operation. We define the integral representation of this operation to be (Equation 4.79).

$$\mathbf{g}(t) = \int_{T_0}^{T_f} \mathbf{Q}_\mathbf{r}(t, \tau)\mathbf{s}(\tau)\, d\tau, \qquad T_0 \leqq t \leqq T_f. \tag{6.7}$$

Operating upon $\mathbf{s}(t)$ with $\mathbf{Q}_\mathbf{r}(t, \tau)$ to find $\mathbf{g}(t)$ is equivalent to solving the integral by means of our differential equation approach. It is easy to show that the inverse kernel satisfies the following integral equation in two variables*:

$$\int_{T_0}^{T_f} \mathbf{K}_\mathbf{r}(t, v)\mathbf{Q}_\mathbf{r}(v, \tau)\, dv = \mathbf{I}\delta(t - \tau), \qquad T_0 < t, \tau < T_f, \tag{6.8}$$

where $\mathbf{I}$ is an r-dimensional identity matrix.

Let us multiply both sides of Equation 6.4 by $\mathbf{Q}_\mathbf{r}(\tau, z)$ and then integrate with respect to τ. This yields

$$\begin{aligned}
&\int_{T_0}^{T_f} \mathbf{K}_\mathbf{x}(t, \tau)\mathbf{C}^T(\tau)\mathbf{Q}_\mathbf{r}(\tau, z)\, d\tau \\
&\quad = \int_{T_0}^{T_f} \mathbf{h}_0(t, v) \int_{T_0}^{T_f} \mathbf{K}_\mathbf{r}(v, \tau)\mathbf{Q}_\mathbf{r}(\tau, z)\, d\tau\, dv \\
&\quad = \int_{T_0}^{T_f} \mathbf{h}_0(t, v)\mathbf{I}\delta(v - z)\, dv = \mathbf{h}_0(t, z), \qquad T_0 \leqq t, z \leqq T_f.
\end{aligned} \tag{6.9}$$

We are not directly interested in the impulse of the optimum estimator. What we really want to find is the estimate $\hat{\mathbf{x}}(t)$, which is the

* We should observe that the inverse kernel can be shown to be symmetric, i.e., $\mathbf{Q}_\mathbf{r}(t, \tau) = \mathbf{Q}_\mathbf{r}^T(\tau, t)$; therefore, we can define it as a pre- or postmultiplier operator.

output of the estimator. We can obtain this by substituting Equation 6.9 into Equation 6.2,

$$\hat{\mathbf{x}}(t) = \int_{T_0}^{T_f} \mathbf{K}_{\mathbf{x}}(t, \tau)\mathbf{C}^T(\tau)\left[\int_{T_0}^{T_f} \mathbf{Q}_{\mathbf{r}}(\tau, z)\mathbf{r}(z)\, dz\right] d\tau, \qquad T_0 \leqq t \leqq T_f. \tag{6.10}$$

Thus, the optimum estimate is the result of two integral operations. We now want to show how we can reduce Equation 6.10 to two differential equations with an associated set of boundary conditions. The estimate $\hat{\mathbf{x}}(t)$ is specified implicitly by their solution.

Let us define the term in brackets in Equation 6.10 as $\mathbf{g}_{\mathbf{r}}(\tau)$, so that we have

$$\mathbf{g}_{\mathbf{r}}(\tau) = \int_{T_0}^{T_f} \mathbf{Q}_{\mathbf{r}}(\tau, z)\mathbf{r}(z)\, dz, \qquad T_0 \leqq \tau \leqq T_f. \tag{6.11}$$

Substituting this into Equation 6.9 gives us

$$\hat{\mathbf{x}}(t) = \int_{T_0}^{T_f} \mathbf{K}_{\mathbf{x}}(t, \tau)\mathbf{C}^T(\tau)\mathbf{g}_{\mathbf{r}}(\tau)\, d\tau, \qquad T_0 \leqq t \leqq T_f. \tag{6.12}$$

Observe that Equations 6.10 and 6.11 are integral operations of the type encountered in Chapters 4 and 2, respectively. Consequently, we can convert each into two vector differential equations with an associated set of boundary conditions.

From our previous discussion, $\mathbf{g}_{\mathbf{r}}(\tau)$ is the solution to the inhomogeneous Fredholm integral equation when the signal $\mathbf{s}(t)$ is replaced by $\mathbf{r}(t)$. From Chapter 4, we have

$$\mathbf{g}_{\mathbf{r}}(\tau) = \mathbf{R}^{-1}(\tau)(\mathbf{r}(\tau) - \mathbf{C}(\tau)\boldsymbol{\xi}(\tau)), \tag{6.13}$$

where $\boldsymbol{\xi}(\tau)$ is the solution to the differential equations (Equations 4.14 and 4.16),

$$\frac{d\boldsymbol{\xi}(\tau)}{d\tau} = \mathbf{F}(\tau)\boldsymbol{\xi}(t) + \mathbf{G}(\tau)\mathbf{Q}\mathbf{G}^T(\tau)\boldsymbol{\eta}_1(\tau), \qquad T_0 \leqq \tau \leqq T_f, \tag{6.14}$$

$$\frac{d\boldsymbol{\eta}_1(\tau)}{d\tau} = \mathbf{C}^T(\tau)\mathbf{R}^{-1}(\tau)\mathbf{C}(\tau)\boldsymbol{\xi}(\tau) - \mathbf{F}^T(\tau)\boldsymbol{\eta}_1(\tau) - \mathbf{C}^T(\tau)\mathbf{R}^{-1}(\tau)\mathbf{r}(\tau),$$
$$T_0 \leqq \tau \leqq T_f. \tag{6.15}$$

The boundary conditions are (Equations 4.18 and 4.19)

$$\boldsymbol{\xi}(T_0) = \mathbf{P}_0\, \boldsymbol{\eta}_1(T_0), \tag{6.16}$$

$$\boldsymbol{\eta}_1(T_f) = \mathbf{0}. \tag{6.17}$$

In Chapter 2 we found that the integral operation given by Equation 6.12 also has a differential equation representation. If in Equation 2.30 we set

$$\mathbf{f}(t) = \mathbf{g}_r(t), \qquad T_0 \leqq t \leqq T_f, \tag{6.18}$$

and then substitute Equation 6.13 for $\mathbf{g}_r(t)$, we find that $\mathbf{x}(t)$ can be found by solving the differential equations (Equations 2.40–2.42)

$$\frac{d\hat{\mathbf{x}}(t)}{dt} = \mathbf{F}(t)\hat{\mathbf{x}}(t) + \mathbf{G}(t)\mathbf{Q}\mathbf{G}^T(t)\boldsymbol{\eta}_2(t), \qquad T_0 \leqq t \leqq T_f, \tag{6.19}$$

$$\begin{aligned}\frac{d\boldsymbol{\eta}_2(t)}{dt} &= -\mathbf{C}^T(t)\mathbf{g}_r(t) - \mathbf{F}^T(t)\boldsymbol{\eta}_2(t) \\ &= \mathbf{C}^T(t)\mathbf{R}^{-1}(t)\mathbf{C}(t)\boldsymbol{\xi}(t) - \mathbf{F}^T(t)\boldsymbol{\eta}_2(t) - \mathbf{C}^T(t)\mathbf{R}^{-1}(t)\mathbf{r}(t), \\ &\qquad T_0 \leqq t \leqq T_f.\end{aligned} \tag{6.20}$$

The boundary conditions for these equations (Equations 2.43 and 2.45) are

$$\hat{\mathbf{x}}(T_0) = \mathbf{P}_0\,\boldsymbol{\eta}_2(T_0), \tag{6.21}$$

$$\boldsymbol{\eta}_2(T_f) = 0. \tag{6.22}$$

Upon a first inspection it appears that we need to solve four coupled vector differential equations with their associated boundary conditions. However, if we examine Equations 6.15 and 6.20, we find that $\boldsymbol{\eta}_1(t)$ and $\boldsymbol{\eta}_2(t)$ satisfy the same differential equation. Since both equations have the same boundary condition at $t = T_f$ (Equations 6.17 and 6.22), they must have identical solutions. Therefore, we have

$$\boldsymbol{\eta}_1(t) = \boldsymbol{\eta}_2(t) \triangleq \mathbf{p}(t), \qquad T_0 \leqq t \leqq T_f. \tag{6.23}$$

By replacing $\boldsymbol{\eta}_1(t)$ and $\boldsymbol{\eta}_2(t)$ by $\mathbf{p}(t)$ in Equations 6.13, 6.15, 6.18, and 6.20, we see that $\boldsymbol{\xi}(t)$ and $\hat{\mathbf{x}}(t)$ satisfy the same differential equations (Equations 6.14 and 6.19) and have the same boundary conditions (Equations 6.16 and 6.21).

Therefore, we must also have

$$\hat{\mathbf{x}}(t) = \boldsymbol{\xi}(t), \qquad T_0 \leqq t \leqq T_f. \tag{6.24}$$

Consequently, we have shown that two of the four differential equations are redundant.

We finally obtain the state variable representation of the optimum linear smoother. The optimum estimate $\hat{\mathbf{x}}(t)$ satisfies the differential equations

$$\frac{d\hat{\mathbf{x}}(t)}{dt} = \mathbf{F}(t)\hat{\mathbf{x}}(t) + \mathbf{G}(t)\mathbf{Q}\mathbf{G}^T(t)\mathbf{p}(t), \qquad T_0 \leqq t \leqq T_f, \tag{6.25}$$

$$\frac{d\mathbf{p}(t)}{dt} = \mathbf{C}^T(t)\mathbf{R}^{-1}(t)\mathbf{C}(t)\hat{\mathbf{x}}(t) - \mathbf{F}^T(t)\mathbf{p}(t) - \mathbf{C}^T(t)\mathbf{R}^{-1}(t)\mathbf{r}(t), \qquad T_0 \leqq t \leqq T_f, \tag{6.26}$$

where we impose the boundary conditions

$$\hat{\mathbf{x}}(T_0) = \mathbf{P}_0\,\mathbf{p}(T_0), \tag{6.27}$$

$$\mathbf{p}(T_f) = \mathbf{0}. \tag{6.28}$$

The smoother realization specified by Equations 6.25 to 6.28 is well known. It was first derived by Bryson and Frazier in Reference 15 by assuming Gaussian statistics and then using a variational approach to maximize the *a posteriori* probability of the state vector.

When we compare these equations with those in Chapter 4 which specified our solution to the inhomogeneous Fredholm integral equation, we observe that they are identical in form. The major difference is that our input is now a random process $\mathbf{r}(t)$, whereas before we had a known signal $\mathbf{s}(t)$. The result of this observation is that the solution methods developed in Section 4.3 are also applicable to solving the above estimation equations for the smoother. (In fact, the methods presented there were originally developed in the literature for solving these estimator equations. We have used the above identity of form to solve the equation we derived in Chapter 4.) In order to make use of these methods, it is obvious that one must identify $\hat{\mathbf{x}}(t)$ with $\boldsymbol{\xi}(t)$ and $\mathbf{p}(t)$ with $\boldsymbol{\eta}(t)$.

Since we need the results in the next section, it is useful to relate the smoothing structure derived above to the realizable filter structure. To do this we review some of the relationships that we derived in Chapter 4. If we make the identity suggested above, the variable $\boldsymbol{\xi}_r(t)$ is easily seen to correspond to the realizable filter estimate, $\hat{\mathbf{x}}_r(t)$. It is well known, or it can be seen from Equation 4.44, that $\hat{\mathbf{x}}_r(t)$ satisfies the equation

$$\frac{d\hat{\mathbf{x}}_r(t)}{dt} = \mathbf{F}(t)\hat{\mathbf{x}}_r(t) + \boldsymbol{\Sigma}(t \mid t)\mathbf{C}^T(t)\mathbf{R}^{-1}(t)(\mathbf{r}(t) - \mathbf{C}(t)\hat{\mathbf{x}}_r(t)), \qquad T_0 < t, \tag{6.29}$$

where $\Sigma(t\,|\,t)$ satisfies the variance equation (Equation 4.39).

$$\frac{d\Sigma(t\,|\,t)}{dt} = \mathbf{F}(t)\Sigma(t\,|\,t) + \mathbf{F}^T(t)\Sigma(t\,|\,t) - \Sigma(t\,|\,t)\mathbf{C}^T(t)\mathbf{R}^{-1}(t)\mathbf{C}(t)\Sigma(t\,|\,t) + \mathbf{G}(t)\mathbf{Q}\mathbf{G}^T(t). \tag{6.30}$$

We have not demonstrated directly that $\Sigma(t\,|\,t)$ is indeed the covariance matrix of the realizable filter. It is straightforward to do so using the realizable filter theory; therefore, we omit it.

The smoother estimate $\hat{\mathbf{x}}(t)$ is related to the realizable estimate $\hat{\mathbf{x}}_r(t)$ in two ways. Quite obviously, the estimates correspond at the end of the interval as stated by Equation 4.42

$$\hat{\mathbf{x}}(t)\Big|_{t=T_f} = \hat{\mathbf{x}}_r(T_f). \tag{6.31}$$

In addition we have the important relationship throughout the interval as expressed by Equation 4.48

$$\Sigma(t\,|\,t)\mathbf{p}(t) = \hat{\mathbf{x}}(t) - \hat{\mathbf{x}}_r(t), \qquad T_0 \leqq t \leqq T_f. \tag{6.32}$$

We often use this relationship in the remainder of the chapter.

There is one important contrast between the two structures. In the realizable filter, the covariance of error, $\Sigma(t\,|\,t)$, was implicit in the filter structure. In the smoother, the corresponding covariance $\Sigma(t\,|\,T_f)$ is not. Deriving an equation for this covariance is the topic of our next section.

6.2 Covariance of Error for the Optimum Smoother

In this section we derive three matrix differential equations, each of which specifies the performance of the optimal smoother. Since this is a rather long section, we pause briefly to outline our development.

Let us briefly outline our derivation. First we derive a differential equation for the covariance of $\mathbf{p}(t)$. To do this we need to employ a result regarding realizable whitening filters as discussed by Collins.[18] We next relate the covariance of $\mathbf{p}(t)$ to the covariance of the error for the smoother. Finally, we derive an equation whose solution involves both these covariances. This last equation has the advantages of having constant coefficients whereas the previous two did not.

From Equation 6.26 we have that $\mathbf{p}(t)$ satisfies the differential equation

$$\frac{d\mathbf{p}(t)}{dt} = \mathbf{C}^T(t)\mathbf{R}^{-1}(t)\mathbf{C}(t)\hat{\mathbf{x}}(t) - \mathbf{F}^T(t)\mathbf{p}(t) - \mathbf{C}^T(t)\mathbf{R}^{-1}(t)\mathbf{r}(t), \qquad T_0 \leqq t \leqq T_f. \tag{6.33}$$

Substituting Equation 6.32, we obtain

$$\frac{d\mathbf{p}(t)}{dt} = -(\mathbf{F}^T(t) - \mathbf{C}^T(t)\mathbf{R}^{-1}(t)\mathbf{C}(t)\mathbf{\Sigma}(t \mid t))\mathbf{p}(t) - \mathbf{C}^T(t)\mathbf{R}^{-1}(t) \times (\mathbf{r}(t) - \mathbf{C}^T(t)\hat{\mathbf{x}}_r(t)) \qquad T_0 \leqq t \leqq T_f, \tag{6.34}$$

where $\hat{\mathbf{x}}_r(t)$ is the realizable filter estimate. Following Collins we now show that

$$\mathbf{w}'(t) = \mathbf{r}(t) - \mathbf{C}(t)\hat{\mathbf{x}}_r(t) \tag{6.35}$$

is a white process.

We have

$$\begin{aligned} E[\mathbf{w}'(t)\mathbf{w}'^T(\tau)] &= E[(\mathbf{r}(t) - \mathbf{C}(t)\hat{\mathbf{x}}_r(t))(\mathbf{r}(\tau) - \mathbf{C}(\tau)\hat{\mathbf{x}}_r(\tau))^T] \\ &= E[\mathbf{w}(t) - \mathbf{C}(t)\boldsymbol{\varepsilon}_r(t)(\mathbf{w}(\tau) - \mathbf{C}(\tau)\boldsymbol{\varepsilon}_r(\tau))^T], \qquad T_0 < t, \tau, \end{aligned} \tag{6.36}$$

where $\boldsymbol{\varepsilon}_r(t)$ is the realizable filter error given by

$$\boldsymbol{\varepsilon}_r(t) = \hat{\mathbf{x}}_r(t) - \mathbf{x}(t). \tag{6.37}$$

Expanding this expression term by term, we obtain

$$\begin{aligned} E[\mathbf{w}'(t)\mathbf{w}'^T(\tau)] &= \mathbf{R}(t)\delta(t-\tau) - \mathbf{C}(t)E[\boldsymbol{\varepsilon}_r(t)\mathbf{w}^T(\tau)] - E[\mathbf{w}(t)\boldsymbol{\varepsilon}_r^T(\tau)]\mathbf{C}^T(\tau) \\ &\quad + \mathbf{C}(t)E[\boldsymbol{\varepsilon}_r(t)\boldsymbol{\varepsilon}_r^T(\tau)]\mathbf{C}^T(\tau). \end{aligned} \tag{6.38}$$

We now evaluate the last three terms separately. To do this we need to consider the realizable error process $\boldsymbol{\varepsilon}_r(t)$. If we substitute the differential equations for $\hat{\mathbf{x}}_r(t)$ and $\mathbf{x}(t)$, as given by Equation 4.44 and Equation 2.1, respectively, we have that

$$\begin{aligned} \dot{\boldsymbol{\varepsilon}}_r(t) &= (\mathbf{F}(t) - \mathbf{\Sigma}(t \mid t)\mathbf{C}^T(t)\mathbf{R}^{-1}(t)\mathbf{C}(t))\boldsymbol{\varepsilon}_r(t) \\ &\quad + \mathbf{\Sigma}(t \mid t)\mathbf{C}^T(t)\mathbf{R}^{-1}(t)\mathbf{w}(t) - \mathbf{G}(t)\mathbf{u}(t). \end{aligned} \tag{6.39}$$

The initial condition is an independent zero mean random vector $\boldsymbol{\varepsilon}_r(T_0)$ with covariance $\mathbf{P}_0$. We can write the solution to the above in terms of the transition matrix for the differential equation

$$\boldsymbol{\varepsilon}_r(t) = \boldsymbol{\Phi}(t, T_0)\boldsymbol{\varepsilon}_r(T_0) + \int_{T_0}^{t} \boldsymbol{\Phi}(t, t')(\boldsymbol{\Sigma}(t' \mid t')\mathbf{C}^T(t')\mathbf{R}^{-1}(t')\mathbf{w}(t') - \mathbf{G}(t')\mathbf{u}(t'))\, dt', \qquad T_0 < t, \quad (6.40)$$

where $\boldsymbol{\Phi}(t, \tau)$ is the transition matrix associated with

$$(\mathbf{F}(t) - \boldsymbol{\Sigma}(t \mid t)\mathbf{C}^T(t)\mathbf{R}^{-1}(t)\mathbf{C}(t)).$$

We note that $\boldsymbol{\varepsilon}_r(t)$ has orthogonal increments. Evaluating the expectations in Equation 6.38 in a manner similar to Equation 2.13, we obtain

$$E[\boldsymbol{\varepsilon}_r(t)\mathbf{w}^T(\tau)] = \begin{cases} \boldsymbol{\Phi}(t, \tau)\boldsymbol{\Sigma}(\tau \mid \tau)\mathbf{C}^T(\tau), & T_0 < \tau < t, \\ \frac{1}{2}\boldsymbol{\Sigma}(t \mid t)\mathbf{C}^T(t), & \tau = t, \\ \mathbf{0}, & \tau > t, \end{cases} \qquad (6.41)$$

and

$$E[\boldsymbol{\varepsilon}_r(t)\boldsymbol{\varepsilon}_r^T(\tau)] = \begin{cases} \boldsymbol{\Phi}(t, \tau)\boldsymbol{\Sigma}(\tau \mid \tau), & T_0 < \tau \leqq t, \\ \boldsymbol{\Sigma}(t \mid t)\boldsymbol{\Phi}^T(\tau, t), & \tau > t. \end{cases} \qquad (6.42)$$

When we substitute these equations into Equation 6.38, we have

$$E[\mathbf{w}'(t)\mathbf{w}'^T(\tau)] = \mathbf{R}(t)\delta(t - \tau), \qquad (6.43)$$

Consequently, $\mathbf{w}'(t)$ is a white process.

We now integrate Equation 6.34 using its transition matrix. Applying the terminal boundary conditions on $\mathbf{p}(t)$, we have

$$\mathbf{p}(t) = \int_{t}^{T_f} \boldsymbol{\Phi}'(t, t')\mathbf{C}^T(t')\mathbf{R}^{-1}(t')\mathbf{w}'(t')\, dt', \qquad (6.44)$$

where $\boldsymbol{\Phi}'(t, \tau)$ is the transition matrix associated with

$$-(\mathbf{F}^T(t) - \mathbf{C}^T(t)\mathbf{R}^{-1}(t)\mathbf{C}(t)\boldsymbol{\Sigma}(t \mid t)).$$

(Since the transition used in Equation 6.39 and Equation 6.34 is for adjoint systems, we have the relation $\boldsymbol{\Phi}'(t, \tau) = \boldsymbol{\Phi}^T(\tau, t)$.) We now obtain, using Equations 6.44 and 6.43,

$$E[\mathbf{p}(t)\mathbf{p}^T(t)] \triangleq \boldsymbol{\Pi}(t \mid T_f) = \int_{t}^{T_f} \boldsymbol{\Phi}'(t, \tau)\mathbf{C}^T(\tau)\mathbf{R}^{-1}(\tau)\mathbf{C}(\tau)\boldsymbol{\Phi}'^T(t, \tau)\, d\tau, \qquad T_0 \leqq t \leqq T_f, \quad (6.45)$$

and

$$E[\mathbf{p}(t)\mathbf{p}^T(\tau)] = \begin{cases} \mathbf{\Phi}'(t, \tau)\mathbf{\Pi}(\tau \mid T_f), & \tau \geq t, \\ \mathbf{\Pi}(t \mid T_f)\mathbf{\Phi}'^T(\tau, t), & t > \tau, \ T_0 \leqq t \leqq T_f. \end{cases} \quad (6.46)$$

The integral of Equation 6.45 is not the most convenient form for finding $\mathbf{\Pi}(t \mid T_f)$. If we differentiate it using the properties of transition matrices, we obtain

$$\begin{aligned} \frac{d\mathbf{\Pi}(t \mid T_f)}{dt} &= -(\mathbf{F}(t) - \mathbf{\Sigma}(t \mid t)\mathbf{C}^T(t)\mathbf{R}^{-1}(t)\mathbf{C}(t))^T\mathbf{\Pi}(t \mid T_f) \\ &\quad - \mathbf{\Pi}(t \mid T_f)(\mathbf{F}(t) - \mathbf{\Sigma}(t \mid t)\mathbf{C}^T(t)\mathbf{R}^{-1}(t)\mathbf{C}(t)) \\ &\quad - \mathbf{C}^T(t)\mathbf{R}^{-1}(t)\mathbf{C}(t). \end{aligned} \quad (6.47)$$

From Equation 6.28 the boundary condition is obviously

$$\mathbf{\Pi}(T_f \mid T_f) = \mathbf{0}. \quad (6.48)$$

Note that the realizable filter covariance enters as a coefficient in Equation 6.47, which implies that it has time-varying coefficients even for constant-parameter systems. We also note that the coefficient matrices involve the negative of the realizable filter coefficient matrix; consequently, 6.47 is usually stable when integrated backwards from the end point T_f just as we noted for a similar equation in Chapter 4. Consequently, by solving Equation 6.47 and using Equation 6.46 we can determine the covariance $E[\mathbf{p}(t)\mathbf{p}^T(\tau)]$ at any two points in the observation interval.

We now want to determine the covariance of the smoother, $\mathbf{\Sigma}(t \mid T_f)$. To do this we need two important results. The first concerns the cross correlation between the error process of the smoother $\boldsymbol{\varepsilon}(t) = \hat{\mathbf{x}}(t) - \mathbf{x}(t)$ and the process $\mathbf{p}(t)$. By the classical arguments of the orthogonality principle, one can easily show that the received signal $\mathbf{r}(\tau)$ and the error $\boldsymbol{\varepsilon}(t)$ are uncorrelated processes for all $t, \tau \in [T_0, T_f]$.* As given by Equation 6.25 and 6.28 the random process $\mathbf{p}(\tau)$ is the result of a linear operation upon the received signal; consequently, it is also uncorrelated with $\boldsymbol{\varepsilon}(t)$ for all $t, \tau \in [T_0, T_f]$, or

$$E[\boldsymbol{\varepsilon}(t)\mathbf{p}^T(\tau)] = \mathbf{0}, \qquad T_0 \leqq t, \tau \leqq T_f. \quad (6.49)$$

Next we consider Equation 6.32. Subtracting and then adding $\mathbf{x}(t)$, we can write this in the form

$$\boldsymbol{\varepsilon}(t) - \boldsymbol{\varepsilon}_r(t) = \mathbf{\Sigma}(t \mid t()\mathbf{p}(t), \qquad T_0 \leqq t \leqq T_f, \quad (6.50)$$

* Ref. 46, pp. 390–399.

or

$$\boldsymbol{\varepsilon}(t) - \boldsymbol{\Sigma}(t \mid t)\mathbf{p}(t) = \boldsymbol{\varepsilon}_r(t), \qquad T_0 \leqq t \leqq T_f. \tag{6.51}$$

To find $E[\boldsymbol{\varepsilon}(t)\boldsymbol{\varepsilon}^T(\tau)]$, we simply postmultiply Equation 6.41 by its transpose evaluated at τ and then take the expected value of the result using Equation 6.49. We then obtain

$$E[\boldsymbol{\varepsilon}(t)\boldsymbol{\varepsilon}^T(\tau)] + \boldsymbol{\Sigma}(t \mid t)E[\mathbf{p}(t)\mathbf{p}^T(\tau)]\boldsymbol{\Sigma}(\tau \mid \tau) = E[\boldsymbol{\varepsilon}_r(t)\boldsymbol{\varepsilon}_r^T(\tau)]. \tag{6.52}$$

We can evaluate each of these terms individually using Equations 6.46 and 6.42. The particular case of most interest is when $t = \tau$. Here we can drive a differential equation for

$$E[\boldsymbol{\varepsilon}(t)\boldsymbol{\varepsilon}^T(t)] = \boldsymbol{\Sigma}(t \mid \mathbf{T}_f).$$

We have

$$\boldsymbol{\Sigma}(t \mid T_f) = \boldsymbol{\Sigma}(t \mid t) - \boldsymbol{\Sigma}(t \mid t)\boldsymbol{\Pi}(t \mid T_f)\boldsymbol{\Sigma}(t \mid t). \tag{6.53}$$

Straightforward differentiation, substitution of Equations 6.30, 6.47, and 6.52 itself yields the differential equation

$$\begin{aligned}\frac{d\boldsymbol{\Sigma}(t \mid T_f)}{dt} = {} & (\mathbf{F}(t) + \mathbf{G}(t)\mathbf{Q}\mathbf{G}^T(t)\boldsymbol{\Sigma}^{-1}(t \mid t))\boldsymbol{\Sigma}(t \mid T_f) \\ & + \boldsymbol{\Sigma}(t \mid T_f)(\mathbf{F}(t) + \mathbf{G}(t)\mathbf{Q}\mathbf{G}^T(t)\boldsymbol{\Sigma}^{-1}(t \mid t))^T\boldsymbol{\Sigma}(t \mid T_f) \\ & - \mathbf{G}(t)\mathbf{Q}\mathbf{G}^T(t), \qquad T_0 \leqq t \leqq T_f.\end{aligned} \tag{6.54}$$

Several comments are in order here. This formula was first derived by Rauch, Tung, and Striebel using a discrete time analysis and then passing to a continuous limit.[49] The covariance of error for the realizable estimate $\boldsymbol{\Sigma}(t \mid t)$ enters Equation 6.53 in two ways. Obviously, its inverse appears as part of the coefficient terms. In addition, it supplies the required boundary condition at $t = T_f$ for

$$\boldsymbol{\Sigma}(t \mid T_f)\Big|_{t=T_f} = \boldsymbol{\Sigma}(t \mid t)\Big|_{t=T_f}. \tag{6.55}$$

Consequently, we can solve Equation 6.54 backwards in time from $t = T_f$. We can reach the same conclusions regarding the stability of the integration as we did earlier for Equation 6.48 when integrated backwards from $t = T_f$, the solution to Equation 6.54 is in general stable. Consequently, obtaining solutions for long time intervals does not cause any numerical difficulties when done this way. If the realizable filter covariance is already available, it may be preferable to use Equation

6.47 in conjunction with Equation 6.53 to find $\boldsymbol{\Sigma}(t\,|\,T_f)$, since a continuous inversion of $\boldsymbol{\Sigma}(t\,|\,t)$ is not required, as it is in Equation 6.54.

The first two differential equations that we have derived are not well suited for finding analytic solutions for $\boldsymbol{\Sigma}(t\,|\,T_f)$. Even in the case of constant-parameter systems, the matrix differential equations have time-varying coefficients. We can eliminate this difficulty by considering a $2n \times 2n$ matrix differential equation rather than an $n \times n$ one as we have considered before. To do this we need to combine these two forms in an appropriate manner.

One can show by straightforward manipulation of Equations 6.54, 6.47, and 6.51 that the following matrix differential equation is satisfied.

$$\frac{d\mathbf{P}(t\,|\,T_f)}{dt} = \mathbf{W}(t)\mathbf{P}(t\,|\,T_f) + \mathbf{P}(t\,|\,T_f)\mathbf{W}^T(t)$$
$$+ \begin{bmatrix} \mathbf{G}(t)\mathbf{Q}\mathbf{G}^T(t) & \mathbf{0} \\ \mathbf{0} & \mathbf{C}^T(t)\mathbf{R}^{-1}(t)\mathbf{C}(t) \end{bmatrix}, \qquad T_0 \leqq t \leqq T_f, \tag{6.56a}$$

where $\mathbf{W}(t)$ is defined by Equation 4.23.

$$\mathbf{P}(t\,|\,T_f) = \begin{bmatrix} \boldsymbol{\Sigma}(t\,|\,T_f) & -\boldsymbol{\Pi}(t\,|\,T_f)\boldsymbol{\Sigma}(t\,|\,t) \\ -\boldsymbol{\Sigma}(t\,|\,t)\boldsymbol{\Pi}(t\,|\,T_f) & -\boldsymbol{\Pi}(t\,|\,T_f) \end{bmatrix}. \tag{6.56b}$$

We can specify the boundary condition at $t = T_f$ by using Equations 6.48 and 6.55.

$$\mathbf{P}(T_f\,|\,T_f) = \begin{bmatrix} \boldsymbol{\Sigma}(T_f\,|\,T_f) & \mathbf{0} \\ \mathbf{0} & \mathbf{0} \end{bmatrix}, \tag{6.57}$$

where $\boldsymbol{\Sigma}(T_f\,|\,T_f)$ is the realizable filter error at T_f. Consequently we can solve Equation 6.56a backwards over the interval using this boundary condition. This is analogous to the second method for solving the inhomogeneous integral equation that we developed in Chapter 4.

If we extend the concept developed there, it is easy to find an integral representation for the solution to Equation 6.55. We have

$$\mathbf{P}(t\,|\,T_f) = \boldsymbol{\Psi}(t, T_f)\begin{bmatrix} \boldsymbol{\Sigma}(T_f\,|\,T_f) & \mathbf{0} \\ \mathbf{0} & \mathbf{0} \end{bmatrix}\boldsymbol{\Psi}^T(t, T_f)$$
$$-\int_t^{T_f} \boldsymbol{\Psi}(t, \tau)\begin{bmatrix} \mathbf{G}(\tau)\mathbf{Q}\mathbf{G}^T(\tau) & \mathbf{0} \\ \mathbf{0} & \mathbf{C}^T(\tau)\mathbf{R}^{-1}(\tau)\mathbf{C}(\tau) \end{bmatrix}\boldsymbol{\Psi}^T(t, \tau)\,d\tau,$$
$$T_0 \leqq t \leqq T_f, \tag{6.58}$$

where $\boldsymbol{\Psi}(t, \tau)$ is the transition matrix for $\mathbf{W}(t)$.

We should point out that solving for $\mathbf{\Sigma}(t \mid T_f)$ using this form does have certain advantages when the system parameters are constant. In this case, the coefficient matrix $\mathbf{W}$ is a constant matrix, which allows us to see the matrix exponential. This is certainly an analytic advantage. However, we should remember that we have a larger-dimensional set of equations with which to deal. Finally, we observe that this form is not well suited for finding numerical solutions of $\mathbf{\Sigma}(t \mid T_f)$. Since it involves $\mathbf{W}(t)$ as a coefficient matrix, it introduces virtually the same stability problems that we had for a second method in Chapter 4.

We also point out that several errors have appeared in the literature regarding the smoother performance; therefore, one should be somewhat careful with this issue.[15,49] (See Reference 10 for a complete discussion.*)

This completes our discussion of the performance of the optimal smoother. Let us now illustrate it with several examples.

6.3 Examples of the Optimal Smoother Performance

In the last section we analyzed the performance, or covariance of error, of the optimal smoother. In this section we shall illustrate this performance by considering several examples. First, we work two examples for first-order systems analytically. We apply Equations 6.56 and 6.57 to do this. Then we consider the analysis of a second-order system by numerically integrating Equation 6.47 and then applying Equation 6.53.

Example 6.1 Covariance of Error for a Wiener Process

In this first example we shall find the covariance of error for a Wiener process that is received in the presence of additive white noise. The system for generating this process is described in Equations 2.23 and 2.24. We repeat the system matrices here for convenience.

$$\begin{aligned} F &= 0, \\ G &= 1, \\ Q &= 1, \\ C &= \mu, \\ P_0 &= 0. \end{aligned} \tag{6.59}$$

* These principally concern the matrix $\mathbf{\Pi}(t \mid T_f)$. In Ref. 15 the elements of $P(t \,.\, T_f)$ are derived and interpreted incorrectly. In Ref. 49 the equivalent of Equation 6.53 is incorrect, in addition to indicating a correlation between $\mathbf{p}(t)$ and $\boldsymbol{\epsilon}(\tau)$.

In addition we assume that the spectral height of the additive white noise is σ and the observation interval is $[0, T]$; i.e., $T_0 = 0$ and $T_f = T$. Fortunately, we have available many of the required results from previous examples which concern the process. As a result, we can find the solutions rather quickly.

The first step is to find the matrix $\mathbf{W}(t)$ and its associated transition matrix $\mathbf{\Psi}(t, \tau)$. We did this in Chapter 4 and the results are indicated by Equations 4.56 and 4.61. (We need to substitute $t - \tau$ for t since the system has constant parameters.) Next we need to find $\Sigma(T \mid T)$. We do this by using the partitions of $\mathbf{\Psi}(t, \tau)$ as indicated by Equation 4.36. This yields

$$\Sigma(T \mid T) = \left[\frac{\mu^2}{\sigma}\right]^{-1/2} \tanh\left(\left[\frac{\mu^2}{\sigma}\right]^{1/2} T\right). \tag{6.60}$$

Consequently, Equation 6.58 becomes

$$\mathbf{P}(t \mid T) = \mathbf{\Psi}(t, T)\begin{bmatrix} \left[\frac{\mu^2}{\sigma}\right]^{-1/2} \tanh\left(\left[\frac{\mu^2}{\sigma}\right]^{1/2} T\right) & 0 \\ 0 & 0 \end{bmatrix}\mathbf{\Psi}^T(t, T)$$

$$-\int_t^T \mathbf{\Psi}(t, \tau)\begin{bmatrix} 1 & 0 \\ 0 & \dfrac{\mu^2}{\sigma} \end{bmatrix}\mathbf{\Psi}^T(t, \tau)\, d\tau, \qquad 0 < t < T, \tag{6.61}$$

where $\mathbf{\Psi}(t, \tau)$ is as defined by Equation 4.61 as just indicated. Separating out the upper left partition for $\Sigma(t \mid T)$, we find, after some straightforward manipulation,

$$\Sigma(t \mid T) = \left[\frac{\mu^2}{\sigma}\right]^{1/2} \frac{\cosh\left(\left[\frac{\mu^2}{\sigma}\right]^{1/2}(T - t)\right)\sinh\left(\left[\frac{\mu^2}{\sigma}\right]^{1/2} t\right)}{\cosh\left(\left[\frac{\mu^2}{\sigma}\right]^{1/2} T\right)},$$

$$0 \leqq t \leqq T. \tag{6.62}$$

Example 6.2 Covariance of Error for a First-Order Process

Let us now consider the performance for a random process generated by a system with a pole at $-k$. The generation of this process is described by Equations 2.25 and 2.26. The state matrices are repeated here for convenience.

$$\begin{aligned} F &= -k, \\ G &= 1, \\ Q &= 2kP, \\ C &= 1, \\ P_0 &= P. \end{aligned} \tag{6.63}$$

Let us consider a slight variation of this representation. Instead of choosing $P_0 = P$ such that the process generated is stationary, let it remain a free variable. In addition, we chose the level of the observation noise to be σ and the observation interval to be $[0, T]$. Again, we have many results available from previous examples.

The matrix $\mathbf{W}(t)$ and its associated transition matrix $\mathbf{\Psi}(t, \tau)$ are given by Equations 4.67 and 4.72, respectively. (Again, we need to substitute $(t - \tau)$ for t.) Next, we need to find $\Sigma(T \mid T)$. By using Equation 4.36 the realizable filter covariance of error $\Sigma(T \mid T)$ for an arbitrary covariance of the initial state, P_0, is given by

$$\Sigma(T \mid T) = P\left\{\left[\frac{P_0}{P} + \frac{1}{\Lambda}\left(2 - \frac{P_0}{P}\right)\right]e^{k\Lambda T} + \left[\frac{P_0}{P} - \frac{1}{\Lambda}\left(2 - \frac{P_0}{P}\right)\right]e^{-k\Lambda T}\right\}$$
$$\times\left\{\left[1 + \frac{1}{\Lambda}\left(1 + \frac{P_0}{P}\frac{\Lambda^2 - 1}{2}\right)\right]e^{k\Lambda T} + \left[1 - \frac{1}{\Lambda}\left(1 + \frac{P_0}{P}\frac{\Lambda^2 - 1}{2}\right)\right]e^{k\Lambda T}\right\}^{-1} \tag{6.64a}$$

where

$$\Lambda = \left(1 + \frac{2P}{k\sigma}\right)^{1/2}. \tag{6.64b}$$

For this system, Equation 6.58 becomes

$$\mathbf{P}(t \mid T) = \mathbf{\Psi}(t, T)\begin{bmatrix} \Sigma(T \mid T) & 0 \\ 0 & 0 \end{bmatrix}\mathbf{\Psi}^T(t, T)$$
$$-\int_t^T \mathbf{\Psi}(t, \tau)\begin{bmatrix} 2kP & 0 \\ 0 & \dfrac{1}{\sigma} \end{bmatrix}\mathbf{\Psi}^T(t, \tau)\, d\tau, \qquad 0 \leqq t \leqq T, \tag{6.65}$$

where $\Sigma(T\,|\,T)$ is defined by Equation 6.64a. Taking the upper left portion for $\Sigma(t\,|\,T)$, we find after some straightforward manipulation,

$$\Sigma(t\,|\,T) = \left[\cosh[k\Lambda(T-t)] + \frac{\sinh[k\Lambda(T-t)]}{\Lambda}\right]$$

$$\times \left[\Sigma(T\,|\,T)\cosh[k\Lambda(T-t)] + (\Sigma(T\,|\,T) - 2P)\,\frac{\sinh[k\Lambda(T-t)]}{\Lambda}\right],$$

$$0 \leqq t \leqq T. \quad (6.66)$$

In order to illustrate this result let us consider three choices of P_0.

Case a. $P_0 = P$ (*Stationary Process*)

Here we have chosen P_0 such that we are estimating a stationary process. In this case

$$\Sigma(T\,|\,T) = 2P\left[\frac{(\Lambda+1)e^{k\Lambda T} + (\Lambda-1)e^{-k\Lambda T}}{(\Lambda+1)^2 e^{k\Lambda T} - (\Lambda-1)^2 e^{-k\Lambda T}}\right]. \quad (6.67)$$

If we substitute this equation into Equation 6.66, we can show that

$$\Sigma(t\,|\,T) = \Sigma(T\,|\,T)\frac{\left[\left(\cosh[k\Lambda(T-t)] + \frac{1}{\Lambda}\sinh[k\Lambda(T-t)]\right) \times \left(\cosh[k\Lambda t] + \frac{1}{\Lambda}\sinh[k\Lambda t]\right)\right]}{\left(\cosh[k\Lambda T] + \frac{1}{\Lambda}\sinh[k\Lambda T]\right)},$$

$$0 < t < T. \quad (6.68)$$

Therefore, we see that the covariance of error is symmetric with respect to the midpoint of the interval. We can easily show that they approach the large-time results, i.e., the Wiener filtering results, quite easily. In Figure 6.2 we have plotted Equation 6.68, where we have chosen $k = 1$, $\sigma = 1$, $P = 1$, and $T = 2$.

Case b. $P_0 = 2P/\Lambda + 1$ (*Steady-State Realizable Filtering Error*)

In this case we have chosen the initial covariance such that we do not gain any improvement by realizable filtering, i.e.,

$$\Sigma(T\,|\,T) = P_0 = \frac{2P}{\Lambda+1} \quad (6.69)$$

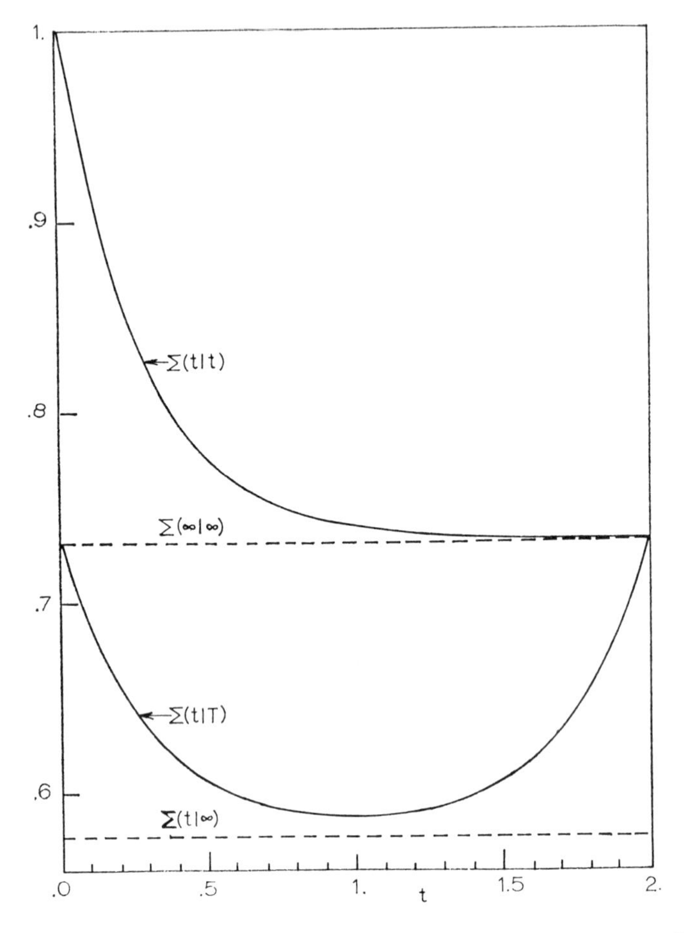

Figure 6.2 $\Sigma(t|t)$ and $\Sigma(t|T)$ for a first-order process, $P_0 = 1$, stationary process.

for all T. Substituting this equation into Equation 6.66 we find

$$\Sigma(t\,|\,T) = \frac{P}{\Lambda} + P\left(\frac{\Lambda - 1}{\Lambda + 1}\right)e^{-2k\Lambda(T-t)}$$

$$= \frac{2P}{\Lambda + 1}\left(\frac{1}{2}\left(1 + \frac{1}{\Lambda}\right) + \frac{1}{2}\left(1 - \frac{1}{\Lambda}\right)e^{-2k\Lambda(T-t)}\right), \qquad 0 \leqq t \leqq T. \tag{6.70}$$

Therefore, we see that the performance approaches a constant as we move backwards from the endpoint. This constant is the unrealizable filter error as we could calculate from the classical theory. The behavior near the end point, i.e., near T, reflects the gain in performance which the realizable filter could attain by allowing some delay before making its estimate. (We develop this concept in detail in the next section.) In Figure 6.3 we have plotted Equation 6.70 for the choice of parameters in the previous figure.

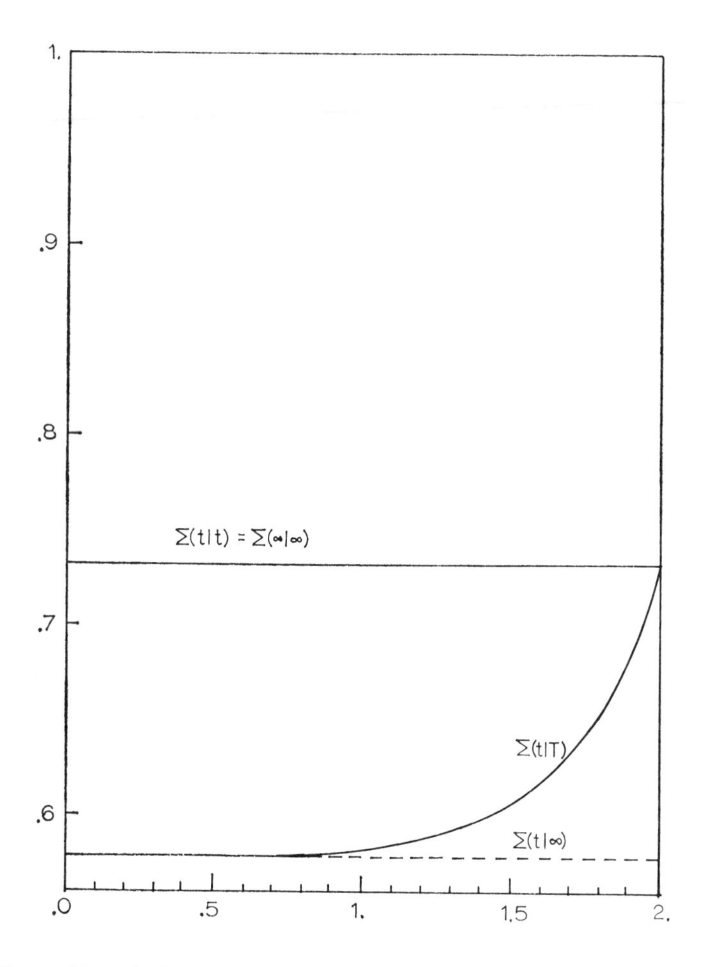

Figure 6.3 $\Sigma(t|t)$ and $\Sigma(t|T)$ for a first-order process, $P_0 = 2/(\Lambda + 1)$, realizable filter in steady state.

Case c. $P_0 = 0$ (*Known Initial State*)

Let us consider the case when we know the initial state exactly, i.e., $P_0 = 0$. This case reflects the extent of correlation time between estimates of the state of the system at different times. For this choice of P_0 we have

$$\Sigma(T \mid T) = \frac{2P \sinh[k\Lambda T]}{\Lambda \cosh[k\Lambda T] + \sinh[k\Lambda T]}. \tag{6.71}$$

When we substitute this into Equation 6.66 we find

$$\Sigma(t \mid T) = \frac{2P}{\Lambda}\left(\frac{\sinh[k\Lambda t]\left(\cosh[k\Lambda(T-t)] + \frac{1}{\Lambda}\sinh[k\Lambda(T-t)]\right)}{\cosh(k\Lambda T) + \frac{1}{\Lambda}\sinh[k\Lambda T]}\right),$$

$$0 \leqq t \leqq T. \tag{6.72}$$

In Figure 6.4 we have plotted $\Sigma(t \mid T)$ for the same choice of parameters. We can see that after 0.6 sec, where it approaches $\Sigma(t \mid \infty)$, the information regarding the initial state is virtually useless in making our estimate. (We point out that we did not take the observation interval quite long enough for this case. There is a plateau in the middle of the interval if the interval is long enough.)

Example 6.3 Covariance of Error for a Second-Order Process

We have analyzed about all the systems one can do analytically in a reasonable amount of time. In order to study higher-dimensioned systems, we use numerical methods. To do this we numerically integrate Equation 6.41 to determine $\mathbf{\Pi}(t \mid T_f)$. Given this function we find $\mathbf{\Sigma}(t \mid T_f)$ by using Equation 6.53. First though, let us consider the system that we wish to study.

We assume that we want to estimate the stationary process $y(t)$ that is described by Equations 2.28 and 2.29. It has a covariance function and a spectrum as illustrated in Figures 2.2a and 2.2b, respectively. Instead of performing the type of analysis done in the previous example, let us consider a variate of a pre-emphasis problem.

Since we have a two-state system, let us see if we can improve our performance by including the second state, or the derivative of the message, in our transmitted signal. In this context, we then have that the

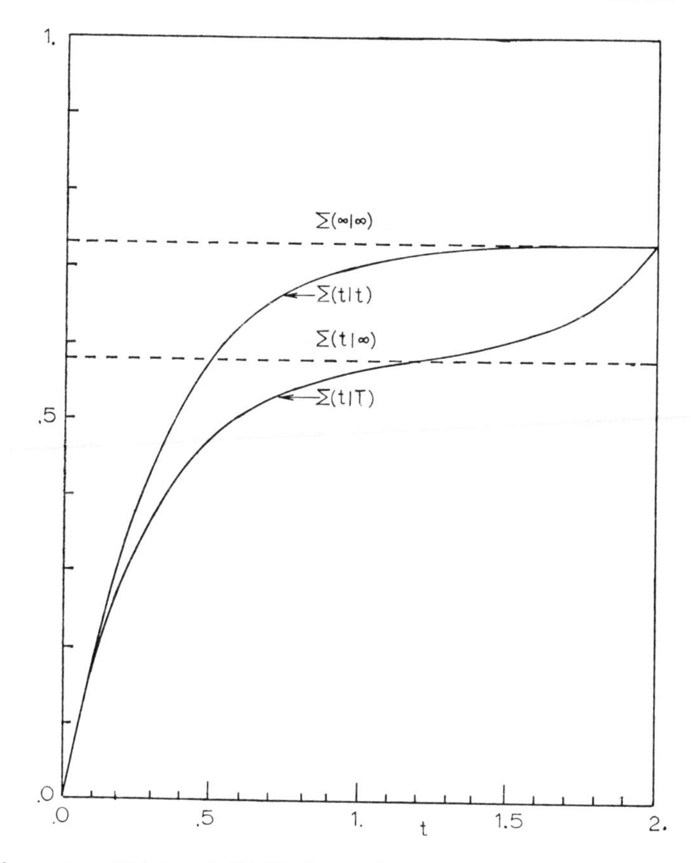

Figure 6.4 $\Sigma(t|t)$ and $\Sigma(t|T)$ for a first-order process, $P_0 = 0$, known initial state.

output of the system is

$$y(t) = \alpha_1 x_1(t) + \alpha_2 x_2(t) = \alpha_1 x_1(t) + \alpha_2 \frac{dx_1(t)}{dt}, \tag{6.73}$$

where we desire to estimate $x_1(t)$ as our message. It would not be a fair comparison to simply add the second state, since this would increase the transmitted power. Let us, therefore, constrain the power to be fixed to its original level of 4. We have

$$E[y^2(t)] = \alpha_1^2 E[x_1^2(t)] + \alpha_2^2 E[x_2^2(t)] = 4\alpha_1^2 + 40\alpha_2^2 = 4. \tag{6.74}$$

$(E[x_1(t)dx_1(t)/dt]) = 0$ for a stationary differentiable random process.) Therefore, we now vary α_1 and α_2 within the constraint of Equation 6.74 to see if it can improve our performance. We also assume that the additive white-noise level is 1.

In summary, the state matrices for our system are

$$\mathbf{F} = \begin{bmatrix} 0 & 1 \\ -10 & -2 \end{bmatrix},$$

$$\mathbf{G} = \begin{bmatrix} 0 \\ 1 \end{bmatrix},$$

$$Q = 160,$$

$$\mathbf{C} = [\alpha_1 \quad \alpha_2],$$

$$R = 1,$$

$$\mathbf{P}_0 = \begin{bmatrix} 4 & 0 \\ 0 & 40 \end{bmatrix},$$

with α_1 and α_2 constrained to be on the ellipse specified by Equation 6.74. In addition we choose the interval length to be 2, or $T_0 = 0$ and $T_f = T = 2$.

If we proceed with our analysis by numerically integrating Equation 6.47 to find $\mathbf{\Sigma}(t \mid T)$ for various values α_1 and α_2 that satisfy Equation 6.74, we generate the curves of Figure 6.5. Here we have plotted $\Sigma_{11}(t \mid T)$ (normalized), the covariance of the message.

We can make several observations regarding these curves. Although we are estimating a stationary process, the curves are not symmetrical about the midpoint of the interval as in Figure 6.2 unless the observed signal contains only one of the states, i.e., $\alpha_1 = 0$, or $\alpha_2 = 0$. Second, we have plotted the curves only for positive values of α_2; if α_2 is negative with respect to α_1, we generate curves that are exactly inverted in time.

Let us also consider the improvement in performance. We have summarized this in Figure 6.6. In this figure we have plotted $\Sigma_{11}(T \mid T)$ (normalized), the realizable estimate at the end point of the interval, and $\Sigma_{11}(T/2 \mid T)$, the smoother performance of the midpoint of the interval. These points are very close to their asymptotic limits, $\Sigma_{11}(\infty \mid \infty)$ and $\Sigma_{11}(t \mid \infty)$. First, we see that transmitting some of the derivative in the signal does not improve the smoother performance. However, it can either degrade or improve the realizable estimate significantly. Choosing $\alpha_2 = 0.2$ improves our performance (over $\alpha_2 = 0$) by approximately 25 per cent, whereas choosing $\alpha_2 = -0.25$ degrades it by 50 per cent. This would indicate from this particular problem that this type of pre-emphasis is not useful when doing smoothing (or filtering with delay), while it can yield significant improvement for realizable filtering.

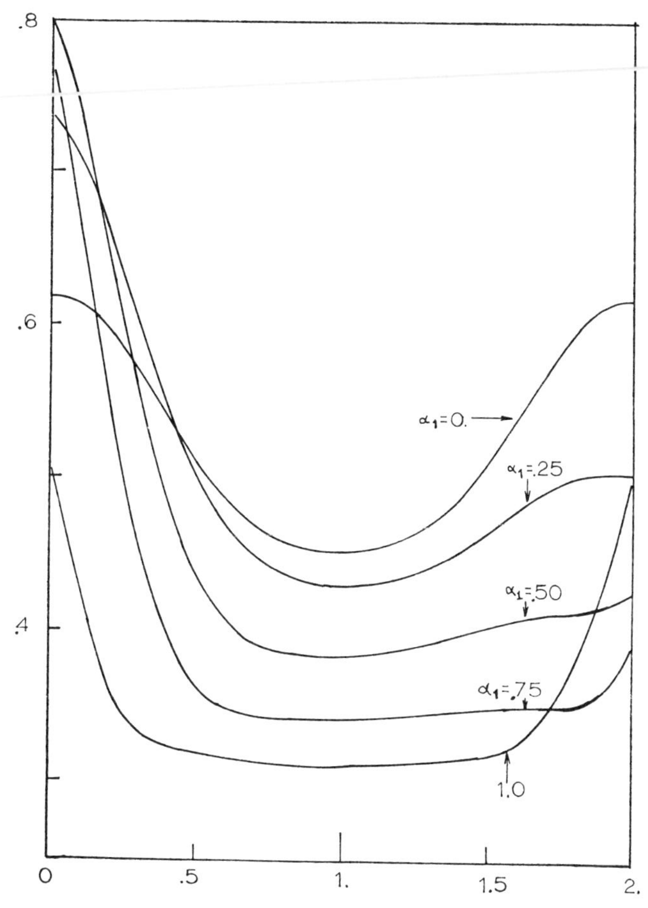

Figure 6.5 $\Sigma_{11}(t|T)$ (normalized) for a second-order process with a transmitted signal $y(t) = \alpha_1 x_1(t) + \alpha_2 x_2(t)$.

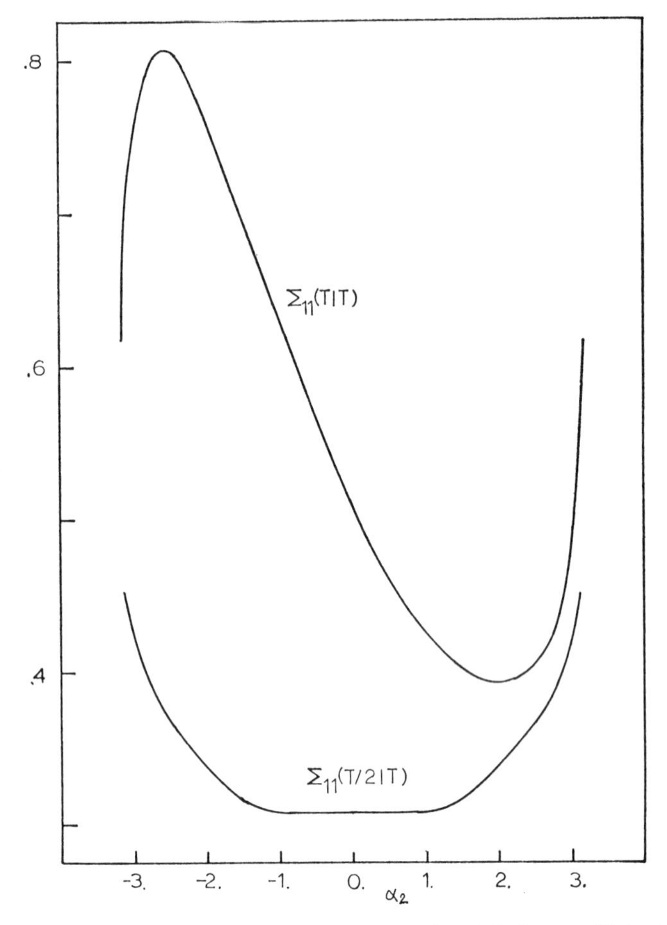

Figure 6.6 $\Sigma_{11}(T|T)$ and $(T/2|T)$ versus α_2 (modulation coefficient of x_2) for a second-order process.

6.4 Filtering with Delay

The optimal smoother uses all the available data, both those in the past and in the future, in making its estimate at a particular point. However, one of the disadvantages of the optimal smoother structure is that it operates over a fixed time interval $[T_0, T_f]$. Consequently, as more data are accumulated, i.e., as T_f increases, the smoothing equations must be solved again if we are to use these added data.

In contrast to this, the realizable filter produces an estimate that evolves as the data are accumulated. It uses, however, only past data in making its estimate, whereas the smoother makes use of both past and future data for its estimate.

The filter realizable with delay combines the advantages of both the realizable filter and the smoother. By allowing a fixed delay before we are required to make our estimate, we can find a filter whose output evolves in time as the data are accumulated, yet is able to take advantage of a limited amount of future data in making its estimate.

Let us discuss the filter structure in more detail. We assume that we have received the signal $\mathbf{r}(t)$ over the interval $[T_0\, T_f]$. We want to estimate the state vector at the point $t = T_f - \Delta$; i.e., find $\hat{\mathbf{x}}(T_f - \Delta)$ with $(T_f - \Delta > T_0)$, where the independent variable in our filter structure is T_f, the end point of the observation interval. We note that, like the realizable filter, the filter with delay is a point estimator, whereas the smoother estimates the signal over an entire interval.

Our approach to finding the filter structure is straightforward.* We use the integral representation specified by solution method 2 of Chapter 4 to find $\hat{\mathbf{x}}(t)$ for the optimal smoother. We then set $t = T_f - \Delta$ in this integral representation. This gives us $\hat{\mathbf{x}}(T_f - \Delta)$ in terms of the received data and the realizable filter. We then differentiate this integral to find a set of differential equations for the desired estimate $\hat{\mathbf{x}}(T_f - \Delta)$. We note that the independent variable for these equations is T_f rather than t, some internal point in a fixed interval.

Let us proceed with our derivation. First we write the smoothing equations Equations 6.25 and 6.26 in augmented matrix form,

$$\frac{d}{dt}\begin{bmatrix}\hat{\mathbf{x}}(t)\\ \mathbf{p}(t)\end{bmatrix} = \mathbf{W}(t)\begin{bmatrix}\hat{\mathbf{x}}(t)\\ \mathbf{p}(t)\end{bmatrix} - \begin{bmatrix}\mathbf{0}\\ \mathbf{C}^T(t)\mathbf{R}^{-1}(t)\mathbf{r}(t)\end{bmatrix}, \qquad T_0 \leqq t \leqq T_f. \tag{6.75}$$

From Equation 4.41 the solution to these equations in an integral form is

$$\begin{bmatrix}\hat{\mathbf{x}}(t)\\ \mathbf{p}(t)\end{bmatrix} = \mathbf{\Psi}(t, T_f)\begin{bmatrix}\hat{\mathbf{x}}_r(T_f)\\ \mathbf{0}\end{bmatrix} + \int_t^{T_f} \mathbf{\Psi}(t, \tau)\begin{bmatrix}\mathbf{0}\\ \mathbf{C}^T(\tau)\mathbf{R}^{-1}(\tau)\mathbf{r}(\tau)\end{bmatrix} d\tau,$$

$$T_0 \leqq t \leqq T_f, \tag{6.76}$$

* This approach to the problem was first used by the author in the 1966 WESCON Proceedings.[6]

where $\hat{\mathbf{x}}_r(T_f)$ is the realizable filter output at the end point of the interval T_f. Let us evaluate Equation 6.75 at $t = T_f - \Delta$, with $T_f - \Delta > T_0$,

$$\begin{bmatrix} \hat{\mathbf{x}}(T_f - \Delta) \\ \mathbf{p}(T_f - \Delta) \end{bmatrix} = \mathbf{\Psi}(T_f - \Delta, T_f) \begin{bmatrix} \hat{\mathbf{x}}_r(T_f) \\ \mathbf{0} \end{bmatrix} + \int_{T_f - \Delta}^{T_f} \mathbf{\Psi}(T_f - \Delta, \tau) \begin{bmatrix} \mathbf{0} \\ \mathbf{C}^T(\tau)\mathbf{R}^{-1}(\tau)\mathbf{r}(\tau) \end{bmatrix} d\tau. \tag{6.77}$$

This is the desired integral representation. We note that the only time variable involved is T_f. Let us now determine a differential equation that Equation 6.77 satisfies, where the independent variable is T_f the end point of an increasing interval rather than t, some internal time in a fixed interval. Differentiating Equation 6.77 we obtain

$$\begin{aligned} \frac{d}{dT_f} \begin{bmatrix} \hat{\mathbf{x}}(T_f - \Delta) \\ \mathbf{p}(T_f - \Delta) \end{bmatrix} &= \frac{d}{dT_f}(\mathbf{\Psi}(T_f - \Delta, T_f)) \begin{bmatrix} \hat{\mathbf{x}}_r(T_f) \\ \mathbf{0} \end{bmatrix} \\ &+ \mathbf{\Psi}(T_f - \Delta, T_f) \begin{bmatrix} \dfrac{d\hat{\mathbf{x}}_r(T_f)}{dT_f} \\ \mathbf{0} \end{bmatrix} \\ &+ \mathbf{\Psi}(T_f - \Delta, T_f) \begin{bmatrix} \mathbf{0} \\ \mathbf{C}^T(T_f)\mathbf{R}^{-1}(T_f)\mathbf{r}(T_f) \end{bmatrix} \\ &- \begin{bmatrix} \mathbf{0} \\ \mathbf{C}^T(T_f - \Delta)\mathbf{R}^{-1}(T_f - \Delta)\mathbf{r}(T_f - \Delta) \end{bmatrix} \\ &+ \int_{T_f - \Delta}^{T_f} \frac{d}{dT_f}(\mathbf{\Psi}(T_f - \Delta, \tau)) \begin{bmatrix} 0 \\ \mathbf{C}^T(\tau)\mathbf{R}^{-1}(\tau)\mathbf{r}(\tau) \end{bmatrix} d\tau. \end{aligned} \tag{6.78}$$

It is a straightforward exercise to show

$$\frac{d}{dT_f}\mathbf{\Psi}(T_f - \Delta, T_f) = \mathbf{W}(T_f - \Delta)\mathbf{\Psi}(T_f - \Delta, T_f) - \mathbf{\Psi}(T_f - \Delta, T_f)\mathbf{W}(T_f). \tag{6.79}$$

We also have

$$\frac{d}{dT_f}\mathbf{\Psi}(T_f - \Delta, \tau) = \mathbf{W}(T_f - \Delta)\mathbf{\Psi}(T_f - \Delta, \tau). \tag{6.80}$$

Let us substitute these two equations plus the expression for $d\hat{\mathbf{x}}_r(T_f)/dT_f$ from Equation 6.29. We obtain

$$\frac{d}{dT_f}\begin{bmatrix}\hat{\mathbf{x}}(T_f-\Delta)\\ \mathbf{p}(T_f-\Delta)\end{bmatrix} = \mathbf{W}(T_f-\Delta)\Bigg\{\mathbf{\Psi}(T_f-\Delta, T_f)\begin{bmatrix}\hat{\mathbf{x}}_r(T_f)\\ \mathbf{0}\end{bmatrix}$$
$$+\int_{T_f-\Delta}^{T_f}\mathbf{\Psi}(T_f-\Delta,\tau)\begin{bmatrix}\mathbf{0}\\ \mathbf{C}^T(\tau)\mathbf{R}^{-1}(\tau)\mathbf{r}(\tau)\end{bmatrix}d\tau\Bigg\}$$
$$+\mathbf{\Psi}(T_f-\Delta, T_f)\Bigg\{-\mathbf{W}(T_f)\begin{bmatrix}\hat{\mathbf{x}}_r(T_f)\\ 0\end{bmatrix}+\begin{bmatrix}\mathbf{0}\\ \mathbf{C}^T(T_f)R^{-1}(T_f)\mathbf{r}(T_f)\end{bmatrix}$$
$$+\begin{bmatrix}(\mathbf{F}T_f)\hat{\mathbf{x}}_r(T_f)+\mathbf{\Sigma}(T_f\,|\,T_f)\mathbf{C}^T(T_f)\mathbf{R}^{-1}(T_f)(\mathbf{r}(T_f)-\mathbf{C}(T_f)\hat{\mathbf{x}}_r(T_f))\\ \mathbf{0}\end{bmatrix}\Bigg\}$$
$$-\begin{bmatrix}\mathbf{0}\\ \mathbf{C}^T(T_f-\Delta)\mathbf{R}^{-1}(T_f-\Delta)\mathbf{r}(T_f-\Delta)\end{bmatrix}. \tag{6.81}$$

First, we identify the term in the first bracket as

$$\begin{bmatrix}\hat{\mathbf{x}}(T_f-\Delta)\\ \mathbf{p}(T_f-\Delta)\end{bmatrix}$$

from Equation 6.77. If we write out the term in the second bracket, we find that it can be written as

$$\begin{bmatrix}\mathbf{\Sigma}(T_f\,|\,T_f)\\ \mathbf{I}\end{bmatrix}\mathbf{C}^T(T_f)\mathbf{R}^{-1}(T_f)(\mathbf{r}(T_f)-\mathbf{C}(T_f)\hat{\mathbf{x}}_r(T_f)).$$

Therefore, we finally obtain the desired differential equation*

$$\frac{d}{dT_f}\begin{bmatrix}\hat{\mathbf{x}}(T_f-\Delta)\\ \mathbf{p}(T_f-\Delta)\end{bmatrix} = \mathbf{W}(T_f-\Delta)\begin{bmatrix}\hat{\mathbf{x}}(T_f-\Delta)\\ \mathbf{p}(T_f-\Delta)\end{bmatrix}$$
$$-\begin{bmatrix}\mathbf{0}\\ \mathbf{C}^T(T_f-\Delta)\mathbf{R}^{-1}(T_f-\Delta)\mathbf{r}(T_f-\Delta)\end{bmatrix}$$
$$+\mathbf{\Psi}(T_f-\Delta, T_f)\begin{bmatrix}\mathbf{\Sigma}(T_f\,|\,T_f)\\ \mathbf{I}\end{bmatrix}\mathbf{C}^T(T_f)\mathbf{R}^{-1}(T_f)(\mathbf{r}(T_f)-C(T_f)\hat{\mathbf{x}}_r(T_f)). \tag{6.82}$$

The only issues that remain are the initial conditions. In order to specify $\hat{\mathbf{x}}(T_0)$ and $\mathbf{p}(T_0)$ we must solve the smoothing equations over the interval $[T_0, T_0+\Delta]$.

* This equation was first obtained by Baggeroer in the 1966 WESCON Proceedings.[6]

Let us now see if we can simplify our structure by eliminating $\mathbf{p}(T_f - \Delta)$. From Equation 6.42 we have

$$\mathbf{p}(T_f - \Delta) = \mathbf{\Sigma}(T_f - \Delta \mid T_f - \Delta)(\hat{\mathbf{x}}(T_f - \Delta) - \hat{\mathbf{x}}_r(T_f - \Delta)). \tag{6.83}$$

Substituting this into Equation 6.82 we find

$$\begin{aligned}\frac{d\hat{\mathbf{x}}(T_f - \Delta)}{dT_f}& \\ = (\mathbf{F}(T_f - \Delta) &+ \mathbf{G}(T_f - \Delta)\mathbf{Q}\mathbf{G}^T(T_f - \Delta)\mathbf{\Sigma}^{-1}(T_f - \Delta \mid T_f - \Delta))\hat{\mathbf{x}}(T_f - \Delta) \\ &- \mathbf{G}(T_f - \Delta)\mathbf{Q}\mathbf{G}^T(T_f - \Delta)\mathbf{\Sigma}^{-1}(T_f - \Delta \mid T_f - \Delta)\,\hat{\mathbf{x}}_r(T_f - \Delta) \\ &+ (\mathbf{\Psi}_{\xi\xi}(T_f - \Delta, T_f)\mathbf{\Sigma}(T_f \mid T_f) + \mathbf{\Psi}_{\xi\eta}(T_f - \Delta, T_f)) \\ &\times \mathbf{C}^T(T_f)\mathbf{R}^{-1}(T_f)(\mathbf{r}(T_f) - \mathbf{C}(T_f)\hat{\mathbf{x}}_r(T_f)).\end{aligned} \tag{6.84}$$

One of the difficulties in implementing Equation 6.84 is the calculation of the coefficient matrix

$$\mathbf{\Psi}_{\xi\xi}(T_f - \Delta, T_f)\Sigma(T_f \mid T_f) + \mathbf{\Psi}_{\xi\eta}(T_f - \Delta, T_f).$$

For constant-parameter system $\mathbf{\Psi}_{\xi\xi}(T_f - \Delta, T_f)$ and $\mathbf{\Psi}_{\xi\eta}(T_f - \Delta, T_f)$ can be computed independently of T_f. Therefore, one need only evaluate $\Sigma(T_f \mid T_f)$, which is already available from the realizable filter structure. For time-varying systems it may be more efficient to compute this coefficient matrix by solving a differential equation for it. It is a straightforward exercise using Equation 6.79 to show that the coefficient matrix satisfies the equation

$$\begin{aligned}\frac{d}{dT_f}&[\mathbf{\Psi}_{\xi\xi}(T_f - \Delta, T_f)\mathbf{\Sigma}(T_f \mid T_f) + \mathbf{\Psi}_{\xi\eta}(T_f - \Delta, T_f)] \\ &= [\mathbf{F}(T_f - \Delta) + \mathbf{G}(T_f - \Delta)\mathbf{Q}\mathbf{G}^T(T_f - \Delta)\mathbf{\Sigma}^{-1}(T_f - \Delta \mid T_f - \Delta)] \\ &\quad \times [\mathbf{\Psi}_{\xi\xi}(T_f - \Delta, T_f)\mathbf{\Sigma}(T_f \mid T_f) + \mathbf{\Psi}_{\xi\eta}(T_f - \Delta, T_f)] \\ &\quad - [\mathbf{\Psi}_{\xi\xi}(T_f - \Delta, T_f)\mathbf{\Sigma}(T_f \mid T_f) + \mathbf{\Psi}_{\xi\eta}(T_f - \Delta, T_f)] \\ &\quad \times [\mathbf{F}(T_f) + \mathbf{G}(T_f)\mathbf{Q}\mathbf{G}^T(T_f)\mathbf{\Sigma}^{-1}(T_f \mid T_f)],\ T_0 + \Delta < T_f.\end{aligned} \tag{6.85}$$

The initial condition follows by setting $T_f = T_0 + \Delta$. Equations 6.84 together with 6.86 are the same as those derived by Meditch,[44] using a discrete time approach. We simply point out that, depending upon the system, we may be able to compute the coefficient matrix more conveniently than by solving a matrix differential equation. The most important aspect of implementing Equation 6.85 is that it appears to be unstable. Indeed, if implemented directly, it would be. To illustrate where the difficulty lies, let us pause a moment in our discussion.

Let us consider a differential equation representation for the linear time-invariant system with an impulse response

$$h(t) = \begin{cases} e^{\beta t}, & 0 < t < \Delta, \quad \beta > 0, \\ 0, & \text{elsewhere.} \end{cases} \tag{6.86}$$

Certainly this system is stable under any realistic criterion. The output of this system is given by

$$y(t) = \int_{t-\Delta}^{t} x(\tau) e^{\beta(t-\tau)}\, d\tau. \tag{6.87}$$

One can easily show that $y(t)$ satisfies the following differential equation:

$$\frac{dy(t)}{dt} = \beta y(t) + x(t) - e^{\beta\Delta} x(t - \Delta). \tag{6.88}$$

Since β is positive, this would indicate an unstable system, which contradicts what we had in Equation 6.86. The difficulty lies in that we are trying to subtract two responses that are in general unstable to yield a stable net response. This is not very feasible to do in a practical sense.

The consequence of our discussion is that Equation 6.84, which has the same form as Equation 6.88, is not suitable for implementation. We should manipulate it into an integral form like Equation 7.86, which has inputs of $\hat{\mathbf{x}}_r(T_f - \Delta)$, $\hat{\mathbf{x}}(T_f)$, and $\mathbf{r}(T_f)$. We should then realize this representation with a tapped delay line.[54] Note that our delay line will have a finite length of Δ; consequently, we can always realize it as closely as desired by decreasing the tap spacing.

This ends our discussion regarding the structure of the filter with delay. Before proceeding with our discussion of its performance, two comments are in order. First, it is straightforward to find a differential equation for $\mathbf{p}(T_f - \Delta)$ as well as $\hat{\mathbf{x}}(T_f - \Delta)$ from Equations 6.82 and 6.83. Second, we emphasize how quickly the filter equations have been derived from the smoother structure by using our technique.

6.5 Performance of the Filter with Delay

In this section we employ the techniques that we have developed to derive a matrix differential equation for $\Sigma(T_f - \Delta \mid T_f)$, the covariance of error for the filter with delay. This equation is important in two respects. First, it tells us how much we gain in performance by allowing the delay and using the more complicated filter structure. Second, for constant-parameter systems we can find how much delay we must allow so as to approach the limiting performance specified by the unrealizable filter. Since we have already derived many of the required results, our derivation is short.

There are two ways in which we can proceed. We can work with Equation 6.54 and derive $\boldsymbol{\Sigma}(T_f - \Delta \mid T_f)$ from it. However, to be consistent with our approach in deriving the filter structure we use Equation 6.56 and separate out the partition for $\boldsymbol{\Sigma}(T_f - \Delta \mid T_f)$. This also has an advantage in identifying some terms in a form that is easy to compute.

To proceed, let us work with Equation 6.56 using the integral representation for the solution of Equation 6.58. Setting $t = T_f - \Delta$ we have for $T_f - \Delta > T_0$

$$\mathbf{P}(T_f - \Delta \mid T_f) = \begin{bmatrix} \boldsymbol{\Sigma}(T_f - \Delta \mid T_f) & -\boldsymbol{\Pi}(T_f - \Delta \mid T_f) \\ & \times \boldsymbol{\Sigma}(T_f - \Delta \mid T_f - \Delta) \\ -\boldsymbol{\Sigma}(T_f - \Delta \mid T_f - \Delta) & -\boldsymbol{\Pi}(T_f - \Delta \mid T_f) \\ \times \boldsymbol{\Pi}(T_f - \Delta \mid T_f) & \end{bmatrix}$$

$$= \boldsymbol{\Psi}(T_f - \Delta, T_f)\begin{bmatrix} \boldsymbol{\Sigma}(T_f \mid T_f) & \mathbf{0} \\ \mathbf{0} & \mathbf{0} \end{bmatrix}\boldsymbol{\Psi}^T(T_f - \Delta, T_f)$$

$$-\int_{T_f - \Delta}^{T_f} \boldsymbol{\Psi}(T_f - \Delta, \tau)\begin{bmatrix} \mathbf{G}(\tau)\mathbf{Q}\mathbf{G}^T(\tau) & \mathbf{0} \\ \mathbf{0} & \mathbf{C}^T(\tau)\mathbf{R}^{-1}(\tau)\mathbf{C}(\tau) \end{bmatrix}\boldsymbol{\Psi}^T(T_f - \Delta, \tau)\, d\tau. \tag{6.89}$$

Let us now differentiate this expression with respect to T_f. We need to use Equations 6.79, 6.80, and 6.29. Doing this we obtain

$$\frac{d}{dT_f}\mathbf{P}(T_f - \Delta \mid T_f)$$

$$= [\mathbf{W}(T_f - \Delta)\boldsymbol{\Psi}(T_f - \Delta, T_f) - \boldsymbol{\Psi}(T_f - \Delta, T_f)\mathbf{W}(T_f)]$$

$$\times \begin{bmatrix} \boldsymbol{\Sigma}(T_f \mid T_f) & \mathbf{0} \\ \mathbf{0} & \mathbf{0} \end{bmatrix}\boldsymbol{\Psi}^T(T_f - \Delta, T_f)$$

$$+ \boldsymbol{\Psi}(T_f - \Delta, T_f)$$

$$
\times \begin{bmatrix} \mathbf{F}(T_f)\mathbf{\Sigma}(T_f \mid T_f) + \mathbf{\Sigma}(T_f \mid T_f)\mathbf{F}^T(T_f) & \\ \quad + \mathbf{G}(T_f)\mathbf{Q}\mathbf{G}^T(T_f) & \mathbf{0} \\ - \mathbf{\Sigma}(T_f \mid T_f)\mathbf{C}^T(T_f)\mathbf{R}^{-1}(T_f)\mathbf{C}(T_f)\mathbf{\Sigma}(T_f \mid T_f) & \\ \mathbf{0} & \mathbf{0} \end{bmatrix}
$$

$$
\times \mathbf{\Psi}^T(T_f - \Delta, T_f)
$$

$$
+ \mathbf{\Psi}(T_f - \Delta, T_f)\begin{bmatrix} \mathbf{\Sigma}(T_f \mid T_f) & \mathbf{0} \\ \mathbf{0} & \mathbf{0} \end{bmatrix}
$$

$$
\times [\mathbf{\Psi}^T(T_f - \Delta, T_f)\mathbf{W}^T(T_f - \Delta) - \mathbf{W}^T(T_f)\mathbf{\Psi}^T(T_f - \Delta, T_f)]
$$

$$
- \mathbf{\Psi}(T_f - \Delta, T_f)
$$

$$
\times \begin{bmatrix} \mathbf{G}(T_f)\mathbf{Q}\mathbf{G}^T(T_f) & \mathbf{0} \\ \mathbf{0} & \mathbf{C}^T(T_f)\mathbf{R}^{-1}(T_f)\mathbf{C}(T_f) \end{bmatrix}
$$

$$
\times \mathbf{\Psi}^T(T_f - \Delta, T_f)
$$

$$
+ \begin{bmatrix} \mathbf{G}(T_f - \Delta)\mathbf{Q}\mathbf{G}^T(T_f - \Delta) & \mathbf{0} \\ \mathbf{0} & \mathbf{C}^T(T_f - \Delta)\mathbf{R}^{-1}(T_f - \Delta)\mathbf{C}(T_f - \Delta) \end{bmatrix}
$$

$$
+ \mathbf{W}(T_f - \Delta)
$$

$$
\times \int_{T_f - \Delta}^{T_f} \mathbf{\Psi}(T_f - \Delta, \tau)\begin{bmatrix} \mathbf{G}(\tau)\mathbf{Q}\mathbf{G}^T(\tau) & \mathbf{0} \\ \mathbf{0} & \mathbf{C}^T(\tau)\mathbf{R}^{-1}(\tau)\mathbf{C}(\tau) \end{bmatrix}\mathbf{\Psi}^T(T_f - \Delta, \tau)\, d\tau
$$

$$
+ \int_{T_f - \Delta}^{T_f} \mathbf{\Psi}(T_f - \Delta, \tau)\begin{bmatrix} \mathbf{G}(\tau)\mathbf{Q}\mathbf{G}^T(\tau) & \mathbf{0} \\ \mathbf{0} & \mathbf{C}^T(\tau)\mathbf{R}^{-1}(\tau)\mathbf{C}(\tau) \end{bmatrix}\mathbf{\Psi}^T(T_f - \Delta, \tau)\, d\tau
$$

$$
\times \mathbf{W}^T(T_f - \Delta). \tag{6.90}
$$

To reduce this equation, we first note that

$$
\mathbf{W}(T_f)\begin{bmatrix} \mathbf{\Sigma}(T_f \mid T_f) & \mathbf{0} \\ \mathbf{0} & \mathbf{0} \end{bmatrix} = \begin{bmatrix} \mathbf{F}(T_f)\mathbf{\Sigma}(T_f \mid T_f) & \mathbf{0} \\ \mathbf{C}^T(T_f)\mathbf{R}^{-1}(T_f)\mathbf{C}(T_f)\mathbf{\Sigma}(T_f \mid T_f) & \mathbf{0} \end{bmatrix}. \tag{6.91}
$$

After we combine terms with common factors and use Equations 6.89 and 6.91, we have

$$
\frac{d\mathbf{P}(T_f - \Delta \mid T_f)}{dT_f}
$$

$$= \mathbf{W}(T_f - \Delta)\mathbf{P}(T_f - \Delta \mid T_f) + \mathbf{P}(T_f - \Delta \mid T_f)\mathbf{W}^T(T_f - \Delta) - \boldsymbol{\Psi}(T_f - \Delta, T_f) \times \begin{bmatrix} \boldsymbol{\Sigma}(T_f \mid T_f)\mathbf{C}^T(T_f)\mathbf{R}^{-1}(T_f) \times \mathbf{C}(T_f)\boldsymbol{\Sigma}(T_f \mid T_f) & \boldsymbol{\Sigma}(T_f \mid T_f)\mathbf{C}^T(T_f) \times \mathbf{R}^{-1}(T_f)\mathbf{C}(T_f) \\ \mathbf{C}^T(T_f)\mathbf{R}^{-1}(T_f) \times \mathbf{C}(T_f)\boldsymbol{\Sigma}(T_f \mid T_f) & \mathbf{C}^T(T_f)\mathbf{R}^{-1}(T_f)\mathbf{C}(T_f) \end{bmatrix} \times \boldsymbol{\Psi}^T(T_f - \Delta, T_f) + \begin{bmatrix} \mathbf{G}(T_f - \Delta)\mathbf{Q} \times \mathbf{G}^T(T_f - \Delta) & \mathbf{0} \\ \mathbf{0} & \mathbf{C}^T(T_f - \Delta)\mathbf{R}^{-1}(T_f - \Delta) \times \mathbf{C}(T_f - \Delta) \end{bmatrix}. \tag{6.92}$$

Simplifying the second term on the right, we have finally

$$\frac{d\mathbf{P}(T_f - \Delta \mid T_f)}{dT_f} = \mathbf{W}(T_f - \Delta)\mathbf{P}(T_f - \Delta \mid T_f) + \mathbf{P}(T_f - \Delta \mid T_f)\mathbf{W}(T_f - \Delta) + \begin{bmatrix} \mathbf{G}(T_f - \Delta)\mathbf{Q}\mathbf{G}^T(T_f - \Delta) & \mathbf{0} \\ \mathbf{0} & \mathbf{C}^T(T_f - \Delta)\mathbf{R}^{-1}(T_f - \Delta)\mathbf{C}(T_f - \Delta) \end{bmatrix} - \boldsymbol{\Psi}(T_f - \Delta, T_f)\begin{bmatrix} \boldsymbol{\Sigma}(T_f \mid T_f) \\ \mathbf{I} \end{bmatrix}\mathbf{C}^T(T_f)\mathbf{R}^{-1}(T_f)\mathbf{C}(T_f) \times [\boldsymbol{\Sigma}(T_f \mid T_f) \quad \mathbf{I}]\boldsymbol{\Psi}^T(T_f - \Delta, T_f). \tag{6.93}$$

We see that with the exception of the added term the final equation is very similar to the corresponding one for the smoother. To find the initial condition $P(T_0 \mid T_0 + \Delta)$, we need to solve one of the form(s) of the smoother performance equations for $\boldsymbol{\Sigma}(T_0 \mid T_0 + \Delta)$ and $\boldsymbol{\Pi}(T_0 \mid T_0 + \Delta)$. Doing this we find

$$\mathbf{P}(T_0 \mid T_0 + \Delta) = \begin{bmatrix} \boldsymbol{\Sigma}(T_0 \mid T_0 + \Delta) & -\boldsymbol{\Pi}(T_0 \mid T_0 + \Delta)\mathbf{P}_0 \\ -\mathbf{P}_0\boldsymbol{\Pi}(T_0 \mid T_0 + \Delta) & -\boldsymbol{\Pi}(T_0 \mid T_0 + \Delta) \end{bmatrix}. \tag{6.94}$$

Let us now consider the upper left partition for $\boldsymbol{\Sigma}(T_f - \Delta \mid T_f)$. Using Equations 6.89 and 6.53 with $t = T_f - \Delta$, we find

$$\begin{aligned}
&\frac{d\boldsymbol{\Sigma}(T_f - \Delta \mid T_f)}{dT_f} \\
&= [\mathbf{F}(T_f - \Delta) + \mathbf{G}(T_f - \Delta)\mathbf{Q}\mathbf{G}^T(T_f - \Delta)\boldsymbol{\Sigma}^{-1}(T_f - \Delta \mid T_f - \Delta)] \\
&\quad \times \boldsymbol{\Sigma}(T_f - \Delta \mid T_f) + \boldsymbol{\Sigma}(T_f - \Delta \mid T_f) \\
&\quad \times [\mathbf{F}(T_f - \Delta) + \mathbf{G}(T_f - \Delta)\mathbf{Q}\mathbf{G}^T(T_f - \Delta)\boldsymbol{\Sigma}^{-1}(T_f - \Delta \mid T_f - \Delta)]^T \\
&\quad + \mathbf{G}^T(T_f - \Delta)\mathbf{Q}\mathbf{G}^T(T_f - \Delta) \\
&\quad - [\boldsymbol{\Psi}_{\xi\xi}(T_f - \Delta, T_f)\boldsymbol{\Sigma}(T_f \mid T_f) + \boldsymbol{\Psi}_{\xi\eta}(T_f - \Delta, T_f)] \\
&\quad \times \mathbf{C}^T(T_f)\mathbf{R}^{-1}(T_f)\mathbf{C}(T_f) \\
&\quad \times [\boldsymbol{\Psi}_{\xi\xi}(T_f - \Delta, T_f)\boldsymbol{\Sigma}(T_f \mid T_f) + \boldsymbol{\Psi}_{\xi\eta}(T_f - \Delta, T_f)]^T. \qquad (6.95)
\end{aligned}$$

To find the initial condition $\boldsymbol{\Sigma}(T_0 \mid T_0 + \Delta)$, we again need to solve the smoother equations over the interval $[T_0, T_0 + \Delta]$. We can again identify the coefficient matrix

$$\boldsymbol{\Psi}_{\xi\xi}(T_f - \Delta \mid T_f)\boldsymbol{\Sigma}(T_f \mid T_f) + \boldsymbol{\Psi}_{\xi\eta}(T_f - \Delta \mid T_f)$$

that we had in the filter structure. As before, we can evaluate this by one of two ways. We finally note that Equation 6.95, together with Equation 6.85, was first derived by Meditch in Reference 44.

One interesting aspect of Equation 6.95 is that it is a linear equation unless $\Delta = 0$, whereupon it becomes the matrix Riccati equation for $\boldsymbol{\Sigma}(T_f \mid T_f)$, as would be expected. The second aspect is that it is unstable when integrated forward. We therefore suggest a backwards integration as we have done in previous problems. To do this one needs to solve the filter variance equation over the entire interval of interest.

6.6 Example of Performance of the Filter with Delay

Let us illustrate the results of the previous section with an example. To do this we numerically integrated Equation 6.95. The coefficient matrix was evaluated by using the matrix exponential.

The example that we study is a one-pole process with the initial covariance matrix chosen such that a stationary process is generated. The equations that describe the generation of the process are Equations 2.25 and 2.26. In Figure 6.7 we have plotted the $\boldsymbol{\Sigma}(T_f - \Delta \mid T_f)$ that results from our numerical integration versus $T_f - \Delta$ for various values

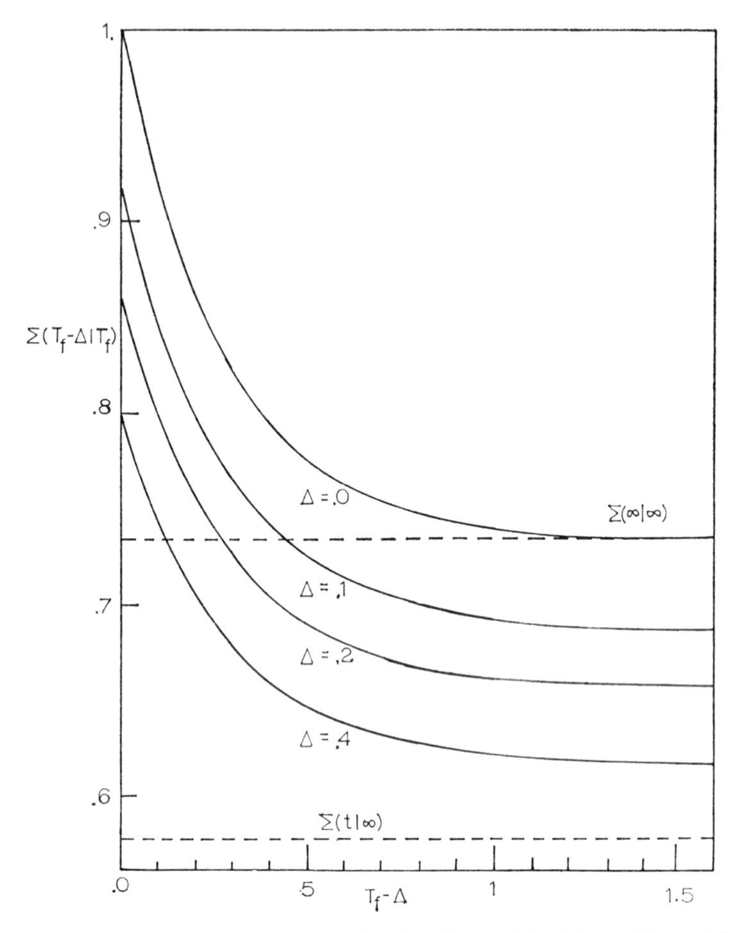

Figure 6.7 Covariance of error for the filter with delay $\Sigma(T_f - \Delta(T_f))$ versus $T_f - \Delta$.

of Δ. The top curve is for $\Delta = 0$, which is the realizable filter covariance as calculated by solving the Ricatti equation; therefore, we can see quickly how much we gain by allowing a delay. We have also indicated the lower limit on the covariance of error, as calculated from the classical Wiener theory.

Several observations should be made. The transient behavior has about the same duration as that of the realizable filter. The curves approach the asymptotic limits very closely over the interval length considered. Finally, we integrated Equation 6.82 forward in time. We have previously indicated that this equation is unstable when integrated in this direction. If we plotted these curves over several more time constants, we could see this instability entering. The curves no longer remain constants as they start to grow exponentially. This oscillation is even more pronounced in the second-order systems that we have studied. This is indicative of the type of behavior we could expect if we integrated the differential equation, Equation 6.82, describing the estimation structure directly.

6.7 Discussion and Conclusions for the Optimal Smoother and Filter with Delay

In this chapter we have extensively discussed linear smoothing and filtering with delay. As we stated in the introduction, we feel that our approach is a unified one in that everything follows from the differential equations for the optimal smoother.

The starting point for our development was the finite-time Wiener-Hopf equation. We presented a method for deriving the smoothing differential equations from this integral equation by using our results for the Fredholm theory developed in Chapters 2 and 4. We then derived several different forms for the differential equations specifying the covariance of error for the smoother.

After working several examples of the smoother performance, we discussed the filter with delay. Both its structure and performance could be derived directly from the smoother results. We also indicated possible instability difficulty in implementing these results. We suggested a way to avoid this difficulty; however, we did not develop the suggestion extensively. We concluded the discussion of filter with delay by presenting an example of its performance for various amounts of delay allowed.

The smoother and/or filter with delay are not specifically limited to the area of estimation theory. For example, quite frequently in problems in radar/sonar the receiver has the optimal smoother as one of its components.[68] Consequently, we can apply our results to realize this part of the receiver. Certainly, this is not the only application; for further discussion we refer to Reference 68.

7 Smoothing and Filtering for Nonlinear Modulation Systems

In the previous chapter we assumed that the generation of the random processes could be described by a linear system with a finite dimensional linear state representation. In this chapter we extend our techniques so as to treat the problems of smoothing and filtering when the observation method is a nonlinear function of the state of the system. To do this, however, we need to make some additional assumptions regarding the statistics of the processes.

With this extension we can represent many modulation schemes and channels of current interest in communications.* See Snyder[61] for a more complete development of this issue. We point out, however, that this is not the most general problem that can be incorporated in the state variable framework, since we still require that our state equation be linear.

Let us outline our procedure. We briefly review our model in order to introduce the notation required for the nonlinearity. We then show how we can use our techniques to convert an integral equation for the optimal smoothed estimate to a pair of differential equations for it. We finally indicate a filtering problem solution when one uses an approximation

* The topic of nonlinear filtering has an extensive literature with the work of Kushner being preeminent.[40] We are concerned here with its relation to the material developed earlier, and not with an extensive discussion of the topic. We refer to Reference 9, for a more complete discussion and bibliography.

technique for converting the smoothing equations described over a fixed time interval to a pair of equations which describes the realizable filter with a moving end point.

7.1 State Variable Model of Linear Systems

In Chapter 2 we introduced the notation required for the generation of random processes by systems with a linear state representation. Let us modify our generation method to allow the observed output, $\mathbf{y}(t)$, to be a nonlinear function of the state vector, $\mathbf{x}(t)$. As before, we require that the internal dynamics of the generation be described by a linear state equation with a random excitation process and random initial conditions. However, here we need to assume that the state vector generated is a vector Gaussian random process. Consequently, we have

$$\frac{d\mathbf{x}(t)}{dt} = \mathbf{F}(t)\mathbf{x}(t) + \mathbf{G}(t)\mathbf{u}(t) \qquad \text{(linear state equation)}, \tag{7.1}$$

$$E[\mathbf{u}(t)\mathbf{u}^T(\tau)] = \mathbf{Q}\,\delta(t-\tau) \qquad \text{(Gaussian white noise)}, \tag{7.2}$$

$$E[\mathbf{x}(T_0)\mathbf{x}^T(T_0) = P_0 \qquad \text{(Gaussian random vector)}, \tag{7.3}$$

where $\mathbf{u}(t)$ is a white Gaussian process, with a "spectral height" of $\mathbf{Q}$, and $\mathbf{x}(T_0)$ is a Gaussian random vector.

The nonlinear aspect of this problem enters in how we observe the state vector. We assume that the output, or observed process, is a continuous, nonlinear, no-memory function of the state vector,

$$\mathbf{y}(t) = \mathbf{s}(\mathbf{x}(t), t), \qquad T_0 \leqq t. \tag{7.4}$$

In summary, we have a nonlinear modulation system, as indicated in Figure 7.1.

Let us also introduce for the gradient of $\mathbf{s}$ with respect to the state vector $\mathbf{x}(t)$

$$\mathbf{C}(\mathbf{x}(t), t) \triangleq \frac{\partial \mathbf{s}(\mathbf{x}(t), t)}{\partial(\mathbf{x}(t))} \triangleq \begin{bmatrix} \frac{\partial s_1(\mathbf{x}(t), t)}{\partial(\mathbf{x}_1(t))} & \frac{\partial s_1(\mathbf{x}(t), t)}{\partial(x_2(t))} & \cdots & \frac{\partial s_1(\mathbf{x}(t), t)}{\partial(x_n(t))} \\ \frac{\partial s_2(\mathbf{x})t(, t)}{\partial(x_1(t))} & & & \\ \vdots & & & \\ \frac{\partial s_m(\mathbf{x}(t), t)}{\partial(x_1(t))} & & & \frac{\partial s_m(\mathbf{x}(t), t)}{\partial(x_n(t))} \end{bmatrix}$$

$$T_0 \leq t. \tag{7.5}$$

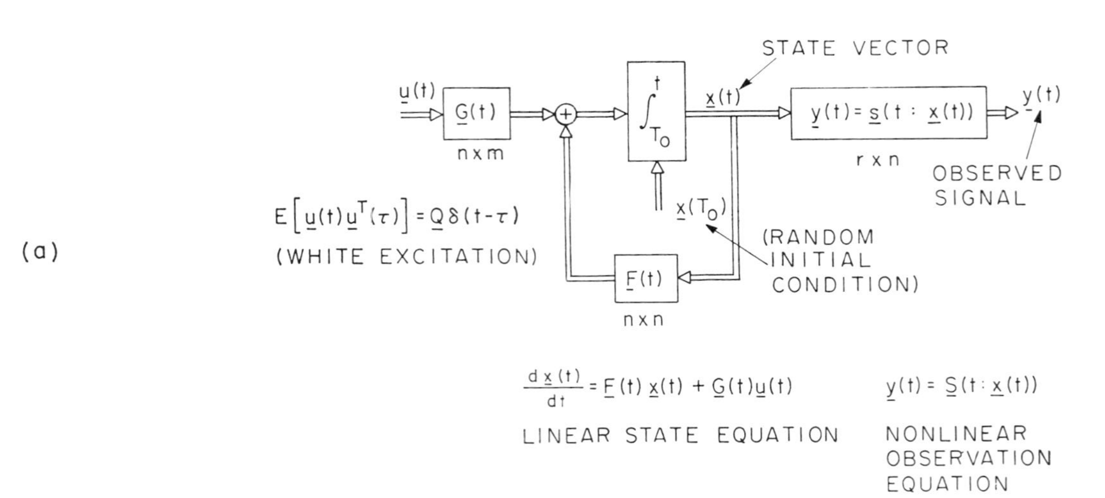

Figure 7.1 (a) State variable model for a system with a nonlinear observation equation.

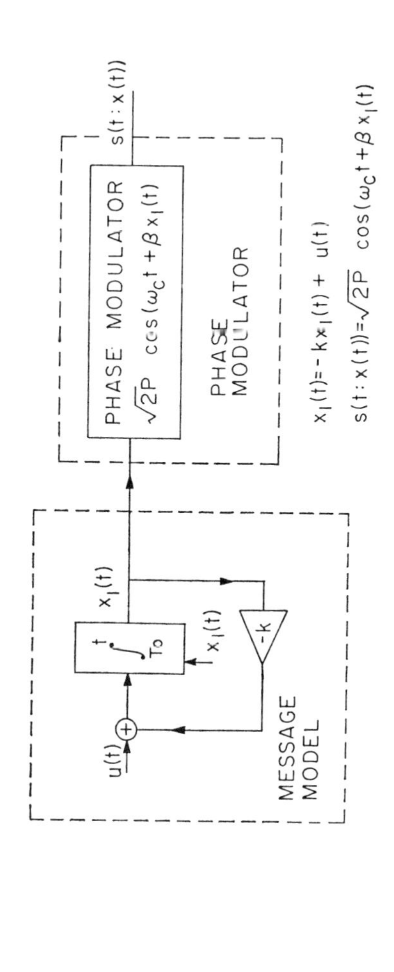

Figure 7.1 (b) State variable model of a PM system.

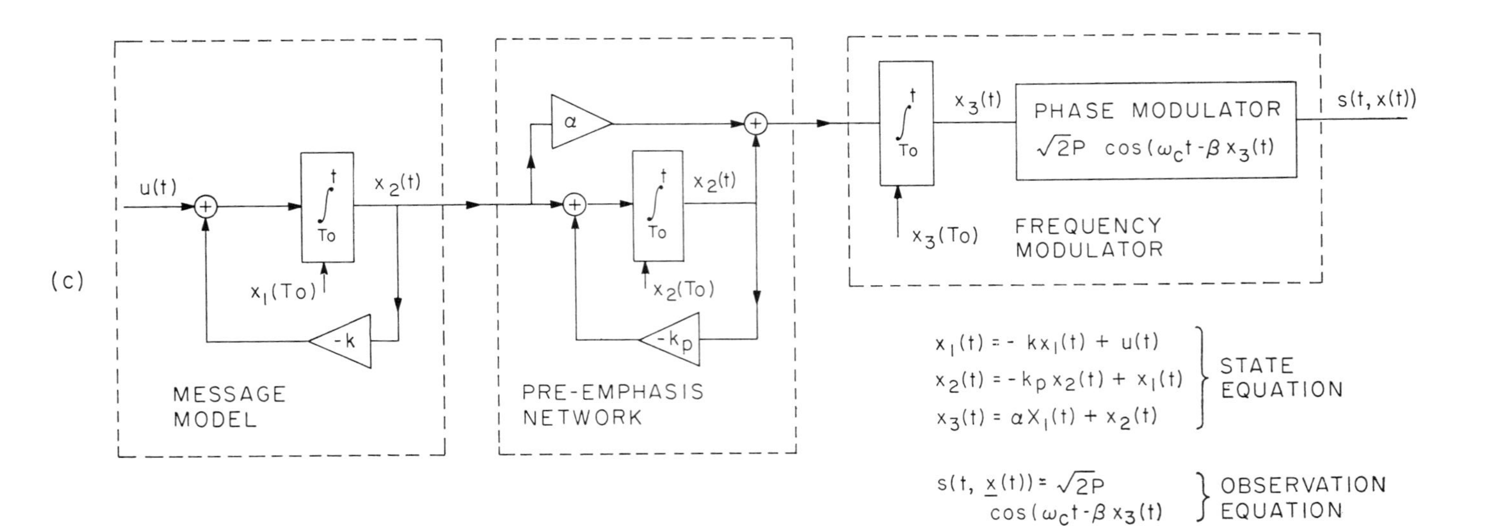

Figure 7.1 (c) State variable model of pre-emphasized FM system.

Therefore, in the special case of linear systems, $\mathbf{C}(\mathbf{x}(t), t)$ is independent of $\mathbf{x}(t)$ and may be identified as the $\mathbf{C}(t)$ that we have been using previously.

We can incorporate certain types of nonlinear memory operation in our structure. If we can interpret the modulation, or observation, operation as the cascade of linear system (with memory) which has a state representation for its dynamics then followed by a nonlinear no-memory operation, we can simply augment the state vector to incorporate the memory operation and then redefine the modulation operation. Probably, the most important application of this is FM modulation, where we interpret it as the cascade of an integrator followed by a phase modulator. Other applications include preemphasized PM or system-identification problems

Again, we still assume that the signal that the receiver observes over the interval $[T_0\ T_f]$ is corrupted by added white noise such that we have

$$\mathbf{r}(t) = \mathbf{s}(\mathbf{x}(t), t) + \mathbf{w}(t), \qquad T_0 \leqq t \leqq T_f, \tag{7.6}$$

where

$$E[\mathbf{w}(t)\mathbf{w}^T(\tau)] = \mathbf{R}(t)\,\delta(t - \tau), \tag{7.7}$$

with $\mathbf{R}(t)$ positive definite. Here we must assume that $\mathbf{w}(t)$ is also Gaussian.

Let us summarize the differences between what we have assumed here and in Chapter 2. First, we allowed the observed signal to be a nonlinear function of the state vector. Second, we have assumed that $\mathbf{u}(t)$ and $\mathbf{w}(t)$ are Gaussian random processes, and $\mathbf{x}(T_0)$ is a Gaussian random vector. In Chapter 2, we made assumptions regarding only their second-order statistics.

One can also consider the problem when the state equation is also nonlinear. However, one needs to use a different approach. In this approach one maximizes the *a posteriori* density directly where the constraint of the state equation is incorporated by using a time-varying Lagrangian multiplier. This was first done in Reference 15.

Finally, we have termed the observation operation as a modulation. This is simply a convenience. One can model a large class of communication systems and channels using this formulation. One simply has a large augmented state vector to incorporate the dynamics of all the systems involved.[61]

7.2 Smoothing for Nonlinear Modulation Systems

We now derive the differential equations and their boundary conditions which implicitly generate the optimal smoothed estimate $\hat{\mathbf{x}}(t)$ when the received signal is generated according to the methods of the previous section. The starting point for derivation is an integral equation for $\hat{\mathbf{x}}(t)$ as derived by Van Trees in Reference 67 using an eigenfunction expansion approach. From Equation 5.160 in this reference, it is necessary that the optimal estimate $\hat{\mathbf{x}}(t)$ satisfy the following integral equation:

$$\hat{\mathbf{x}}(t) = \int_{T_0}^{T_f} \mathbf{K_x}(t, \tau)\mathbf{C}^T(\hat{\mathbf{x}}(\tau), \tau)\mathbf{R}^{-1}(\tau)(\mathbf{r}(\tau) - \mathbf{s}(\hat{\mathbf{x}}(\tau), \tau))\, d\tau, \qquad T_0 \leqq t \leqq T_f. \tag{7.8}$$

We now observe that we have the same type of kernel for the integral operation as we discussed in Section 2.3. The only difference is that before the kernel operated upon $\mathbf{C}^T(\tau)\mathbf{f}(\tau)$, whereas now we have $\mathbf{C}^T(\hat{\mathbf{x}}(\tau), \tau))\mathbf{R}^{-1}(\tau)(\mathbf{r}(\tau) - \mathbf{s}(\hat{\mathbf{x}}(\tau), \tau))$. Consequently, if we want to use the results that we derived in that section, we must examine how this difference of terms affects the derivation. When we do this, we see that the derivation is unaffected. Therefore, in Equation 2.42 we can replace $\mathbf{C}^T(\tau)\mathbf{f}(\tau)$ by $\mathbf{C}^T(\hat{\mathbf{x}}(\tau), \tau)\mathbf{R}^{-1}(\tau)(\mathbf{r}(\tau) - \mathbf{s}(\hat{\mathbf{x}}(\tau), \tau))$ and reduce Equation 7.8 to two differential equations with a set of boundary conditions. The differential equations that describe the optimal smoother for this problem are

$$\frac{d\hat{\mathbf{x}}(t)}{dt} = \mathbf{F}(t)\hat{\mathbf{x}}(t) + \mathbf{G}(t)\mathbf{Q}\mathbf{G}^T(t)\mathbf{p}(t), \qquad T_0 \leqq t \leqq T_f, \tag{7.9}$$

$$\frac{d\mathbf{p}(t)}{dt} = \mathbf{F}^T(t)\mathbf{p}(t) - \mathbf{C}^T(\hat{\mathbf{x}}(t), t)\mathbf{R}^{-1}(t)(\mathbf{r}(t) - \mathbf{s}(\hat{\mathbf{x}}(t), t)), \qquad T_0 \leqq t \leqq T_f. \tag{7.10}$$

The boundary conditions that are imposed are the same. We have

$$\hat{\mathbf{x}}(T_0) = \mathbf{P}_0\, \mathbf{p}(T_0), \tag{7.11}$$

$$\mathbf{p}(T_f) = \mathbf{0}. \tag{7.12}$$

We make two comments regarding Equations 7.9 and 7.10. First, our derivation of these differential equations and boundary conditions from Equation 7.8 is exact. There are no approximations involved. Second, we emphasize that the integral Equation 7.8 is only a necessary condition. It is usually not sufficient; consequently, its solution need not be unique.

Contrasting Equations 7.9 and 7.10 with those derived for the corresponding linear problem in Section 6.1, we see that the equation for $\hat{\mathbf{x}}(t)$

is the same, while the nonlinear aspects of the estimator are introduced in the equation for $\mathbf{p}(t)$.

Now that we have derived the equations for $\hat{\mathbf{x}}(t)$, we must consider methods of solving them. First, let us run through the methods discussed for solving the parallel set of linear equations developed in Section 4.3. The first method employed the superposition principle; therefore, it is not applicable. In the second method we found a complete set of boundary conditions at $t = T_f$ by introducing a function that corresponded to realizable filter output. If we could parallel this we could also solve Equations 7.9 and 7.10 backwards from the end point. This technique is certainly feasible, and we subsequently discuss an approximate solution for the realizable filter. We still may encounter instability problems with this method. The key to the third method was a linear relation between the functions $\hat{\mathbf{x}}(t)$, $\hat{\mathbf{x}}_r(t)$, and $\mathbf{p}(t)$ as given by Equation 4.48. Unfortunately, we do not have such a relationship at the current time. Therefore, only the second method seems at all promising.

Certainly, there do exist other methods of solving nonlinear two-point boundary value problems. Several are discussed in Reference 2. One technique is the method of quasi-linearization. With this technique, the estimation equations are linearized around some *a priori* estimate of the solution. Then these linear equations are solved exactly, by use of the transition matrix associated with this system. This new solution provides the next estimate around which nonlinear equations are linearized. This technique is repeated until a satisfactory convergence criterion has been satisfied. One of the problems, however, is generating the required *a priori* estimate of $\hat{\mathbf{x}}(t)$ and $\mathbf{p}(t)$. A reasonable choice might be to use the estimates found by applying method 2, in effect combining this method with a quasi-linearization approach.

We have seen how the realizable filter is important in solving the smoothing equations. This filter is certainly of much more interest than its use for this application. Let us now discuss how we can use the smoothing equations to find an approximate realization of the realizable filter.

7.3 An Approximate Solution to the Realizable Filter

The fundamental difference between the interval estimator and the realizable filter is the time variable involved. In the smoother the important time variable is the time within the fixed observation interval, whereas in the realizable filter the important time variable is the end point time of the observation interval, which is not fixed but increases continually as the data are accumulated. For the realizable filter we want the estimate at the end point of the observation interval as a function of the length of the interval.

In order to make the transition between the two time variables, one can use the concept of "invariant imbedding." This has been done by Detchmendy and Shridar in Reference 25.* Although we make some minor modifications to avoid some troublesome issues, our derivation essentially parallels theirs.

Let us motivate its use for this particular problem. We consider a sample function of $\mathbf{p}(t)$ near the end point of the observation interval or near $t = T_f$. From Equation 7.12 we have that it vanishes at $t = T_f$. However, at $t = T_f - \Delta T$, we have from Equation 7.10.

$$\mathbf{p}(T_f - \Delta T) = -\left.\frac{d\mathbf{p}(t)}{dt}\right|_{t=T_f} \Delta T$$
$$= \mathbf{C}^T(\hat{\mathbf{x}}(T_f), T_f)\mathbf{R}^{-1}(T_f)(\mathbf{r}(T_f) - \mathbf{s}(\hat{\mathbf{x}}(T_f), T_f))\Delta T \triangleq \Delta\boldsymbol{\eta}. \tag{7.13}$$

Now, we examine the same problem with the same sample functions from a slightly different viewpoint. Let us consider the trajectories for $\hat{\mathbf{x}}(t)$ and $\mathbf{p}(t)$ over the interval $[T_0, T_f - \Delta T]$. We can say that these trajectories solve a second problem defined over this shortened interval. For this problem the initial conditions are the same. However, the end point time is now $T_f - \Delta T$, and $\mathbf{p}(T_f - \Delta T)$ is equal to $\Delta\boldsymbol{\eta}$ as defined in Equation 7.13 instead of being identically zero. We can produce the same trajectory by considering an appropriately chosen nonzero boundary condition for $\mathbf{p}(T_f - \Delta T)$.

This leads us to the following hypothetical question. If we imbed our smoothing problem in a larger class of problems with the boundary condition

$$\mathbf{p}(T_f) = \boldsymbol{\eta} \tag{7.14}$$

for Equation 7.10, how does the solution of Equation 7.9 at T_f depend upon changes in T_f and $\boldsymbol{\eta}$? This question is answered by the invariant imbedding equation, which is a partial differential equation for the solution of Equation 7.9 at T_f as a function of T_f and $\boldsymbol{\eta}$.

First, we sketch its derivation. Let us consider the solutions to Equations 7.9 and 7.10 when we impose the final boundary condition specified by Equation 7.14. We denote these solutions by $\mathbf{x}(t, T_f, \boldsymbol{\eta})$ and $\mathbf{p}(t, T_f, \boldsymbol{\eta})$. We have introduced arguments T_f and $\boldsymbol{\eta}$ to emphasize

* Bellman originally developed the concept of invariant imbedding.[13,14]

that these solutions are dependent upon these parameters. We also point out that we are assuming $\boldsymbol{\eta}$ to be an independent variable. We have

$$\mathbf{x}(t, T_f, \mathbf{0}) = \hat{\mathbf{x}}(t), \qquad T_0 \leqq t \leqq T_f, \tag{7.15}$$

$$\mathbf{p}(T_f, T_f, \boldsymbol{\eta}) = \boldsymbol{\eta}. \tag{7.16}$$

We define

$$\boldsymbol{\xi}(T_f, \boldsymbol{\eta}) \triangleq \mathbf{x}(T_f, T_f, \boldsymbol{\eta}). \tag{7.17}$$

We note

$$\boldsymbol{\xi}(T_f, \mathbf{0}) = \hat{\mathbf{x}}(T_f), \tag{7.18}$$

which is the desired realizable filter estimate at T_f. For convenience we also define the function $\boldsymbol{\Gamma}$ by

$$\begin{aligned}\boldsymbol{\Gamma}(\mathbf{x}(T_f, T_f, \boldsymbol{\eta}), \boldsymbol{\eta}, T_f) &\triangleq \left.\frac{d\mathbf{x}(t, T_f, \boldsymbol{\eta})}{dt}\right|_{t=T_f} \\ &= \mathbf{F}(T_f)\boldsymbol{\xi}(T_f, \boldsymbol{\eta}) + \mathbf{G}(T_f)\mathbf{Q}\mathbf{G}^T(T_f)\boldsymbol{\eta}\end{aligned} \tag{7.19}$$

and the function $\boldsymbol{\Delta}$ by

$$\begin{aligned}\boldsymbol{\Delta}(\mathbf{x}(T_f, T_f, \boldsymbol{\eta}), \boldsymbol{\eta}, T_f) &\triangleq \left.\frac{d\mathbf{p}(t, T_f, \boldsymbol{\eta})}{dt}\right|_{t=T_f} \\ = -\mathbf{F}^T(T_f)\boldsymbol{\eta} &- \mathbf{C}^T(\boldsymbol{\xi}(T_f, \boldsymbol{\eta}), T_f)\mathbf{R}^{-1}(T_f)(\mathbf{r}(T_f) - \mathbf{s}(\boldsymbol{\xi}(T_f, \boldsymbol{\eta}), T_f)),\end{aligned} \tag{7.20}$$

We now determine a partial differential equation for $\boldsymbol{\xi}(T_f, \boldsymbol{\eta})$ in terms of the variables T_f and $\boldsymbol{\eta}$. Let us examine the solutions $\hat{\mathbf{x}}(t, T_f, \boldsymbol{\eta})$ and $\mathbf{p}(t, T_f, \boldsymbol{\eta})$ as illustrated in Figure 7.2. We have from Equation 7.10

$$\begin{aligned}\mathbf{p}(T_f - \Delta T, T_f, \boldsymbol{\eta}) &= \boldsymbol{\eta} - \left.\frac{d\mathbf{p}(t, T_f, \boldsymbol{\eta})}{dt}\right|_{t=T_f} \Delta T \\ &= \boldsymbol{\eta} - \boldsymbol{\Delta}(\hat{\mathbf{x}}(T_f, T_f, \boldsymbol{\eta}), \boldsymbol{\eta}, T_f)\,\Delta T \\ &= \boldsymbol{\eta} - \boldsymbol{\Delta}(\boldsymbol{\xi}(T_f, \boldsymbol{\eta}), \boldsymbol{\eta}, T_f)\,\Delta T \triangleq \boldsymbol{\eta} - \Delta\boldsymbol{\eta}.\end{aligned} \tag{7.21}$$

Now, we can interpret $\mathbf{x}(T_f - \Delta T, T_f, \boldsymbol{\eta})$ as being the solution to a second problem producing the same trajectory over the interval $[T_0, T_f - \Delta T]$

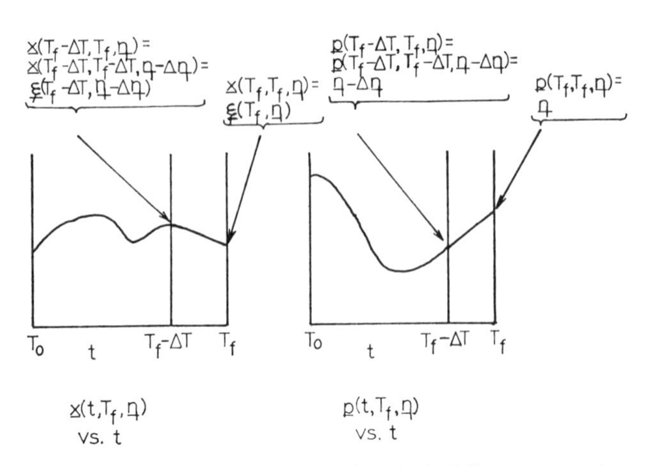

Figure 7.2 Diagram for invariant imbedding argument.

with the boundary condition $\boldsymbol{\eta} - \Delta\boldsymbol{\eta}$, i.e.,

$$\begin{aligned}\mathbf{x}(T_f - \Delta T, T_f, \boldsymbol{\eta}) &= \mathbf{x}(T_f - \Delta T, T_f - \Delta T, \boldsymbol{\eta} - \Delta\boldsymbol{\eta}) \\ &= \boldsymbol{\xi}(T_f - \Delta T, \boldsymbol{\eta} - \Delta\boldsymbol{\eta}). \end{aligned} \tag{7.22}$$

We also have

$$\begin{aligned}\mathbf{x}(T_f - \Delta T, T_f, \boldsymbol{\eta}) &= \mathbf{x}(T_f, T_f, \boldsymbol{\eta}) - \left.\frac{d\mathbf{x}(t, T_f, \boldsymbol{\eta})}{dt}\right|_{t=T_f} \Delta T \\ &= \boldsymbol{\xi}(T_f, \boldsymbol{\eta}) - \boldsymbol{\Gamma}(\boldsymbol{\xi}(T_f, \boldsymbol{\eta}), \boldsymbol{\eta}, T_f)\,\Delta T. \end{aligned} \tag{7.23}$$

Combining Equations 7.22 and 7.23, we find

$$\boldsymbol{\xi}(T_f - \Delta T\ \ \boldsymbol{\eta} - \Delta\boldsymbol{\eta}) = \boldsymbol{\xi}(T_f, \boldsymbol{\eta}) - \boldsymbol{\Gamma}(\boldsymbol{\xi}(T_f, \boldsymbol{\eta}), \boldsymbol{\eta}, T_f)\,\Delta T. \tag{7.24}$$

We also can expand $\boldsymbol{\xi}(T_f, \boldsymbol{\eta})$ in a two-dimensional Taylor series. Doing this, we obtain*

$$\boldsymbol{\xi}(T_f - \Delta T, \eta - \Delta\boldsymbol{\eta}) = \boldsymbol{\xi}(T_f, \boldsymbol{\eta}) - \frac{\partial\boldsymbol{\xi}(T_f, \boldsymbol{\eta})}{\partial T_f}\Delta T - \frac{\partial\boldsymbol{\xi}(T_f, \boldsymbol{\eta})}{\partial\boldsymbol{\eta}}\Delta\boldsymbol{\eta}. \tag{7.25}$$

However, as given by Equation 7.21, we have constrained $\Delta\boldsymbol{\eta}$ to be

$$\Delta\boldsymbol{\eta} = \boldsymbol{\Delta}(\boldsymbol{\xi}(T_f, \boldsymbol{\eta}), \boldsymbol{\eta}, T_f)\,\Delta T. \tag{7.26}$$

* We interpret $\partial\boldsymbol{\xi}(T_f, \boldsymbol{\eta})/\partial\boldsymbol{\eta}$ as in Equation 7.5.

Substituting Equation 7.26 in Equation 7.25, equating the result to Equation 7.24 and dividing by ΔT, we obtain the desired invariant imbedding equation

$$\frac{\partial \boldsymbol{\xi}(T_f, \boldsymbol{\eta})}{\partial T_f} + \frac{\partial \boldsymbol{\xi}(T_f, \boldsymbol{\eta})}{\partial \boldsymbol{\eta}} \cdot \boldsymbol{\Delta}(\boldsymbol{\xi}(T_f, \boldsymbol{\eta}), \boldsymbol{\eta}, T_f) = \boldsymbol{\Gamma}(\boldsymbol{\xi}(T_f, \boldsymbol{\eta}), \boldsymbol{\eta}, T_f). \tag{7.27}$$

This equation relates the value of the solution to Equation 7.9 at $t = T_f$ to changes in T_f and $\boldsymbol{\eta}$, the end point time of the interval and the boundary condition for Equation 7.10 at $t = T_f$.

We now solve Equation 7.27 and evaluate its solution at $\boldsymbol{\eta} = 0$ as prescribed by Equation 7.18 to find the realizable estimate $\mathbf{x}(T_f)$ vs T_f.*

Let us see if we can find a set of ordinary differential equations that generate the solution to Equation 7.27. This equation is a partial differential equation; therefore, we would expect that its solution would require an infinite set of equations. In general, this is true; however, let us try a finite-order approximation. Since we are interested in the solution at $\boldsymbol{\eta} = 0$, let us use a power series expansion in $\boldsymbol{\eta}$. We try a solution of the form

$$\boldsymbol{\xi}(T_f, \boldsymbol{\eta}) = \hat{\mathbf{x}}(T_f) + \mathbf{P}_1(T_f)\boldsymbol{\eta} + \text{terms of } O(|\boldsymbol{\eta}|^3). \tag{7.28}$$

Equation 7.28 implies that we consider explicitly only terms linear in $\boldsymbol{\eta}$.

Let us substitute our trial solution into Equation 7.27. Using Equations 7.17 and 7.18 and expanding terms to first order we obtain

$$\begin{aligned}
&\frac{d\mathbf{x}(T_f)}{dT_f} + \frac{d\mathbf{P}_1(T_f)}{dT_f}\boldsymbol{\eta} + \mathbf{P}_1(T_f)\Big\{-\mathbf{F}^T(T_f)\boldsymbol{\eta} - \mathbf{C}^T(\hat{\mathbf{x}}(T_f), T_f) \\
&\quad \times \mathbf{R}^{-1}(T_f)(\mathbf{r}(T_f) - \mathbf{s}(\hat{\mathbf{x}}(T_f), T_f)) \\
&\quad - \frac{\partial}{\partial \mathbf{x}}(\mathbf{C}^T(\hat{\mathbf{x}}, T_f)\mathbf{R}^{-1}(T_f)(\mathbf{r}(T_f) - \mathbf{s}(\hat{\mathbf{x}}, T_f)))\Big|_{\hat{\mathbf{x}} = \hat{\mathbf{x}}(T_f)} \mathbf{P}_1(T_f)\boldsymbol{\eta}\Big\} \\
&\quad + \text{terms of } O(|\boldsymbol{\eta}|^2) \\
&\quad = \mathbf{F}(T_f)[\hat{\mathbf{x}}(T_f) + \mathbf{P}_1(T_f)\boldsymbol{\eta}] + \mathbf{G}(T_f)\mathbf{Q}\mathbf{G}^T(T_f)\boldsymbol{\eta} \\
&\quad + \text{terms of } O(|\boldsymbol{\eta}|^2).
\end{aligned} \tag{7.29}$$

* We point out that we have made an expansion in terms of ΔT. There is some controversy regarding the significance of the terms in the expansion. With our approach we are certainly involved in this issue. See Ref. 61 for a more complete discussion.

Now we combine terms of the same order in η. We find

$$\left(\frac{d\hat{\mathbf{x}}(T_f)}{dT_f} - \mathbf{F}(T_f)\hat{\mathbf{x}}(T_f) - \mathbf{P}_1(T_f)\mathbf{C}^T(\hat{\mathbf{x}}(T_f), T_f)\mathbf{R}^{-1}(T_f)\right.$$
$$\left.\times (\mathbf{r}(T_f) - \mathbf{s}(\hat{\mathbf{x}}(T_f), T_f))\right)\boldsymbol{\eta}^0 + \left(\frac{d\mathbf{P}_1(T_f)}{dT_f} - \mathbf{P}_1(T_f)\mathbf{F}^T(T_f) - \mathbf{P}_1(T_f)\right.$$
$$\times \frac{\partial}{d\mathbf{x}}(\mathbf{C}^T(\hat{\mathbf{x}}, T_f)\mathbf{R}^{-1}(T_f)(\mathbf{r}(T_f) - \mathbf{s}(\hat{\mathbf{x}}, T_f)))\bigg|_{\hat{\mathbf{x}}=\hat{\mathbf{x}}(T_f)} \mathbf{P}_1(T_f)$$
$$\left. - \mathbf{F}(T_f)\mathbf{P}_1(T_f) - \mathbf{G}(T_f)\mathbf{Q}\mathbf{G}^T(T_f)\right)\boldsymbol{\eta} + \text{terms of } O(|\boldsymbol{\eta}|^2) = 0. \tag{7.30}$$

Demanding a solution for arbitrary $\boldsymbol{\eta}$ gives us a first-order approximation to the realizable filter. For arbitrary $\boldsymbol{\eta}$ the coefficients of each power of $\boldsymbol{\eta}$ must vanish. We find

$$\frac{d\hat{\mathbf{x}}(T_f)}{dT_f} = \mathbf{F}(T_f)\hat{\mathbf{x}}(T_f) + \mathbf{P}_1(T_f)\mathbf{C}^T(\hat{\mathbf{x}}(T_f), T_f)\mathbf{R}^{-1}(T_f)$$
$$\times (\mathbf{r}(T_f) - \mathbf{s}(\hat{\mathbf{x}}(T_f), T_f)), \tag{7.31}$$

$$\frac{d\mathbf{P}_1(T_f)}{dT_f} = \mathbf{F}(T_f)\mathbf{P}_1(T_f) + \mathbf{P}_1(T_f)\mathbf{F}(T_f) + \mathbf{G}(T_f)\mathbf{Q}\mathbf{G}^T(T_f)$$
$$- \mathbf{P}_1(T_f)\frac{\partial}{\partial x}(\mathbf{C}^T(\hat{\mathbf{x}}, T_f))\mathbf{R}^{-1}(T_f)(\mathbf{r}(T_f) - \mathbf{s}(\hat{\mathbf{x}}, T_f))\bigg|_{\hat{\mathbf{x}}=\hat{\mathbf{x}}(T_f)} \mathbf{P}_1(T_f). \tag{7.32}$$

To complete the solution, we need to specify some initial conditions for Equations 7.31 and 7.32. To do this, we set $T_f = T_0$ in Equation 7.28. We have

$$\boldsymbol{\xi}(T_0, \boldsymbol{\eta})\bigg|_{\boldsymbol{\eta}=\mathbf{0}} = \hat{\mathbf{x}}(T_0) = 0, \tag{7.33}$$

since we have assumed zero *a priori* means. We also have that Equation 7.28 must satisfy the condition of Equation 7.11. This implies

$$\mathbf{P}_1(T_0) = \mathbf{P}_0. \tag{7.34}$$

(This also requires that the initial condition for the coefficients in any higher-order expansion must be zero.)

Several comments are in order.

1. Although we have not made an issue about terms of order ΔT in our expansion, we have derived the same approximation as found by Snyder, who approximated the solution to an *a posteriori* Fokker-Planck equation.[61]

2. If the observation method is linear, Equations 7.31 and 7.32 are identical with those describing the Kalman-Bucy filter. This is certainly to be expected. For the linear case, it is easy to show that a first-order expansion yields an exact solution to the invariant imbedding equation.

3. $\mathbf{P}_1(T_f)$ is conditioned upon the received signal; therefore, it cannot be computed *a priori* as in the linear case. We also point out that we have no reason from this method to equate $\mathbf{P}_1(T_f)$ to a conditional covariance matrix. However, in Snyder's approach one can make this identification.

4. Finally, if we want to consider higher-order approximations, we should observe how $\mathbf{P}_1(T_f)$ couples to the estimate. In general, the highest-order terms will also couple both ways, i.e., with the estimate and with the other terms. In addition, the number of elements in the higher-order approximations are going to be large, e.g., on the order of $(NF)^n$, where NF is the state vector dimension and n is the approximation order.

In summary, we have briefly outlined an approach to smoothing and realizable filtering for nonlinear modulation systems. We started with an integral equation that specified a necessary condition for the optimal smoothed estimate. We then demonstrated how some of the techniques that we developed earlier in Chapter 2 could be extended to reduce this integral equation to a pair of nonlinear differential equations for the optimal smoothed estimate. This reduction was an exact procedure; therefore, solving the differential equations is equivalent to solving the integral equations.

These differential equations specified the smoother structure for our problem; however, we were still faced with the issue of solving them. One of the methods suggested employed the realizable filter.

With this motivation in addition to the general desirability of solution for the realizable filter, we introduced the concept of invariant imbedding. This concept used the smoother structure to derive a partial differential equation for the realizable filter estimate. This equation was difficult to implement; therefore, we introduced an approach that allowed us to find an approximate solution which could be implemented conveniently. Using this approach, the filter structure followed directly.

Appendix A Computation of the Exponential Matrix*

For this algorithm, one can demonstrate that for any $R \times R$ matrix A

$$[\mathbf{I}s - \mathbf{A}]^{-1} = \sum_{r=0}^{R-1} \frac{s^{R-(r+1)}}{d(s)} \mathbf{B}_r, \tag{A1}$$

where $d(s)$ is the characteristic polynomial associated with A,

$$d(s) = s^R + d_1 s^{R-1} + \cdots + d_{R-1}s + d_R. \tag{A2}$$

The coefficients d_r and matrices $\mathbf{B}_r$ are determined by

$$\mathbf{B}_0 = \mathbf{I}, \tag{A3a}$$

$$\begin{aligned} d_r &= -\frac{1}{r}\,\mathrm{Tr}[\mathbf{B}_{r-1}\mathbf{A}] \\ \mathbf{B}_r &= \mathbf{B}_{r-1}\mathbf{A} + d_r\mathbf{I}, \qquad r = 1, \ldots, R. \end{aligned} \tag{A3b}$$

The Cayley-Hamilton theorem implies that $\mathbf{B}_R$ is identically zero. Consequently, we have

$$\mathbf{e}^{\mathbf{A}t} = \sum_{r=1}^{R-1} \mathbf{B}_r g_r(t), \tag{A4}$$

where

$$g_r(t) = \mathscr{L}^{-1}\left(\frac{s^{R-(r+1)}}{d(s)}\right) = \sum_{k=0}^{\infty} c_{k+R-(r+1)} \frac{t^k}{k!}, \tag{A5}$$

* This material is based on the Leverrier-Faddeeva algorithm of Ref. 27.

where we have expanded the $g_r(t)$ in a Taylor series about $t = 0$. The coefficients g_r, k can be found using the initial value theorem by recursion

$$c_i = \begin{cases} 0, & 0 \leq i < R - 2, \\ 1, & i = R - 1, \end{cases} \tag{A6}$$

$$c_{i+R} = \sum_{j=1}^{R} d_j c_{R+i-j}, \qquad i = 0, \ldots.$$

Note that all the coefficients are independent of the particular value of t; consequently, they can be precomputed.

One can also nest the sum implied by Equation A5. This gives some improvement in speed and accuracy. If t is much larger than the shortest time constant implied by the roots of the characteristic polynomial $d(s)$ (large $2WT$ product), then it is useful to segment the interval and use the semigroup property of the transition matrix. The accuracy of the various methods has been studied in several places.[43,26,31]

Appendix B Complex Random Process Generation*

All of the waveforms considered in the text were low-pass signals. In many applications, e.g., the signal design problem that we considered, it is useful to be able to extend these concepts to the case of bandpass waveforms. In this section we show how we can do this by using the concept of a complex state variable.

The use of complex notation for representing narrow-band processes and functions is well known. For example, if $y(t)$ is a random process that has a spectrum which is narrow band about a carrier ω_c we can represent it in the form

$$y(t) = y_c(t)\cos(\omega_c t) + y_s(t)\sin(\omega_c t) = \text{Re}[\tilde{y}(t)e^{j\omega_c t}], \tag{B1a}$$

where

$$\tilde{y}(t) = y_c(t) - jy_s(t). \tag{B1b}$$

$y_c(t)$ and $y_s(t)$ are low-pass processes. Under the narrow-band and stationarity assumptions, one can show

$$K_{y_c y_c}(\Delta t) = K_{y_s y_s}(\Delta t) \tag{B2a}$$

and

$$K_{y_c y_s}(\Delta t) = -K_{y_c y_s}(-\Delta t). \tag{B2b}$$

* A more complete discussion regarding complex random processes can be found in Ref. 68, Appendix A. Some of this material also appears in Ref. 8.

Both components have the same covariance, and the cross covariance is an odd function. There are two important points to be made here. We can represent $y(t)$ as sinusoidal modulation of a low-pass complex process $y(t)$, termed the complex envelope. The real and imaginary parts of $y(t)$, $y_c(t)$, and $y_s(t)$, respectively, are random processes that have identical auto covariances, and a very particular form for their cross covariance. This provides a key to our analysis.

The use of complex state variables is purely a notational convenience. It is obvious that all the problems that one wants to consider may be solved by expanding the terms into their real and imaginary components. However, for the problem where the complex notation is applicable, this expansion is too general a formulation and it leads to a needlessly cumbersome description of the processes involved.

If we are to describe random processes by complex notation, we want to have a convenient form for representing the various covariances of the components. Obviously, if we have to enumerate them all individually, we have not gained anything over using the higher-dimensioned representation. The complex notation for describing random processes is applicable when there are two processes; e.g., the quadrature components $y_c(t)$ and $y_s(t)$ of a narrow-band process, which have the same auto covariances and the particular form for the cross covariance between them. We now want to show that we can find a state representation that generates a complex random process that satisfies Equation B2. In addition we want to generalize our concepts so as to include the nonstationary vector process case.

B.1 Random Process Generation with a Complex Notation

In this section we shall develop the theory needed to describe the generation of complex random processes. Let us assume that we have the following state variable description of a linear system:

$$\frac{d\tilde{\mathbf{x}}(t)}{dt} = \tilde{\mathbf{F}}(t)\tilde{\mathbf{x}}(t) + \tilde{\mathbf{G}}(t)\tilde{\mathbf{u}}(t) \qquad \text{(linear state equation)}, \tag{B3a}$$

$$\tilde{\mathbf{y}}(t) = \tilde{\mathbf{C}}(t)\tilde{\mathbf{x}}(t) \qquad \text{(linear observation equation)}, \tag{B3b}$$

where all the coefficient matrices may be complex. In order to describe the generation we make two assumptions regarding the statistics of the driving noise $\tilde{\mathbf{u}}(t)$ and the initial state vector $\tilde{\mathbf{x}}(T_0)$. With these two assumptions we develop the entire theory. Finally, we demonstrate that such representations can indeed be used as a convenience to describe the complex envelope of narrow-band processes by showing that they yield results consistent with Equation B1 and B2 in the stationary case.

First, let us consider the white-noise driving function $\tilde{\mathbf{u}}(t)$. The complex covariance function for the process assuming zero mean is

$$\tilde{\mathbf{K}}_{\tilde{\mathbf{u}}}(t, \tau) = E[\tilde{\mathbf{u}}(t)\tilde{\mathbf{u}}^{+}(\tau)] = \tilde{\mathbf{Q}}\delta(t - \tau), \tag{B4}$$

where we have used the notation

$$\tilde{\mathbf{u}}^{+}(t) = [\tilde{\mathbf{u}}(t)^{*}]^{T}, \tag{B5}$$

i.e., the conjugate transpose. Let us expand this complex covariance in terms of the quadrature components.

$$\begin{aligned}\tilde{\mathbf{K}}_{\tilde{\mathbf{u}}}(t, \tau) &= E[(\mathbf{u}_c(t) - j\mathbf{u}_s(t))(\mathbf{u}_c^T(\tau) + j\mathbf{u}_s(\tau))] \\ &= \mathbf{K}_{\mathbf{u}_c\mathbf{u}_c}(t, \tau) + \mathbf{K}_{\mathbf{u}_s\mathbf{u}_s}(t, \tau) + j\mathbf{K}_{\mathbf{u}_c\mathbf{u}_s}(t, \tau) - j\mathbf{K}_{\mathbf{u}_s\mathbf{u}_c}(t, \tau) \\ &= \tilde{\mathbf{Q}}\ \delta(t - \tau).\end{aligned} \tag{B6}$$

In order that the covariance matrix be a convenient method of representing the covariances and cross covariances of the components of $\tilde{\mathbf{u}}(t)$, we require that

$$\mathbf{K}_{\mathbf{u}_c\mathbf{u}_c}(t, \tau) = \mathbf{K}_{\mathbf{u}_s\mathbf{u}_s}(t, \tau) = \tfrac{1}{2}\ \mathrm{Re}[\tilde{\mathbf{Q}}]\ \delta(t - \tau), \tag{B7a}$$

$$\mathbf{K}_{\mathbf{u}_c\mathbf{u}_s}(t, \tau) = -\mathbf{K}_{\mathbf{u}_s\mathbf{u}_c}(t, \tau) = \tfrac{1}{2}\ \mathrm{Im}[\tilde{\mathbf{Q}}]\ \delta(t - \tau). \tag{B7b}$$

The covariance matrices for the two components are identical nonnegative definite matrices, and the cross covariance matrix is a skew symmetric matrix. This implies that $\tilde{\mathbf{Q}}(t)$ is a Hermitian matrix with a nonnegative definite real part.

We also note that the conjugate operation in the definition of the complex covariance matrix, for under the above assumption we have

$$E[\tilde{\mathbf{u}}(t)\tilde{\mathbf{u}}^{T}(\tau)] = \tilde{\mathbf{Q}}^{1}\delta(t - \tau) = \mathbf{0}. \tag{B8}$$

Quite often, one does not have correlation between the components of $\mathbf{u}(t)$ (i.e., $E[\mathbf{u}_c(t)\mathbf{u}_s^T(t)] = \mathbf{0}$) since any correlation between the components of the state vector may be represented in the coefficient matrices $\tilde{\mathbf{F}}(t)$ and $\tilde{\mathbf{G}}(t)$. In this case, $\tilde{\mathbf{Q}}(t)$ is a real nonnegative definite symmetric matrix. Also, note that, under the assumptions we made, u_{c_i} is uncorrelated with u_{s_i} for all i.

The next issue that we want to consider is the initial conditions. In order that we be consistent with the concept of state, whatever assumptions we make regarding the state vector at the initial time T_0 should be satisfied at an arbitrary time t ($t \geqq T_0$).

First, we assume that $\tilde{\mathbf{x}}(T_0)$ is a complex random vector (we assume zero mean for simplicity). The complex covariance matrix for this random vector is

$$\begin{aligned}\tilde{\mathbf{P}}_0 &= \tilde{\mathbf{K}}_{\tilde{\mathbf{x}}}(T_0, T_0) = E[\tilde{\mathbf{x}}(T_0)\tilde{\mathbf{x}}^+(T_0)] \\ &= \mathbf{K}_{\mathbf{x}_c\mathbf{x}_c}(T_0, T_0) + \mathbf{K}_{\mathbf{x}_s\mathbf{x}_s}(T_0, T_0) \\ &\quad + j\mathbf{K}_{\mathbf{x}_c\mathbf{x}_s}(T_0, T_0) - j\mathbf{K}_{\mathbf{x}_s\mathbf{x}_c}(T_0, T_0).\end{aligned} \tag{B9}$$

The assumptions that we make about the covariance of this random vector are

$$\mathbf{K}_{\mathbf{x}_c\mathbf{x}_c}(T_0, T_0) = \mathbf{K}_{\mathbf{x}_s\mathbf{x}_s}(T_0, T_0) = \tfrac{1}{2}\,\mathrm{Re}[\tilde{\mathbf{P}}_0], \tag{B10a}$$

$$\mathbf{K}_{\mathbf{x}_c\mathbf{x}_s}(T_0, T_0) = -\mathbf{K}_{\mathbf{x}_s\mathbf{x}_c}(T_0, T_0) = \tfrac{1}{2}\,\mathrm{Im}[\tilde{\mathbf{P}}_0]. \tag{B10b}$$

Consequently, the complex covariance matrix of initial condition is a Hermitian matrix with a nonnegative definite real part. We also note that under the above assumptions

$$E[\tilde{\mathbf{x}}(T_0)\tilde{\mathbf{x}}^T(T_0)] = \mathbf{0}. \tag{B11}$$

Let us now consider what these assumptions imply about the covariance of the state vector $\tilde{\mathbf{x}}(t)$ and the observed signal $\tilde{\mathbf{y}}(t)$. Since we can relate the covariance of $\mathbf{y}(t)$ directly to that of the state vector, we consider $\tilde{\mathbf{K}}_{\mathbf{x}}(t, \tau)$ first.

In the study of real state variable random processes one can determine $\tilde{\mathbf{K}}_{\tilde{\mathbf{x}}}(t, \tau)$ in terms of the state equation matrices, the function $\tilde{\mathbf{Q}}$ associated with the covariance of the excitation noise $\tilde{\mathbf{u}}(t)$, and the covariance $\tilde{\mathbf{K}}_{\tilde{\mathbf{x}}}(T_0, T_0)$ of the initial state vector, $\tilde{\mathbf{x}}(T_0)$. The results for complex state variables are exactly parallel. The only change is that the transpose operation is replaced by a conjugate transpose operation. The methods for determining $\tilde{\mathbf{K}}_{\tilde{\mathbf{x}}}(t, \tau)$ are first to find $\tilde{\mathbf{K}}_{\tilde{\mathbf{x}}}(t, t)$ as the solution of a linear differential equation and then use the transition matrix associated with the matrix $\tilde{\mathbf{F}}(t)$ to relate $\tilde{\mathbf{K}}_{\tilde{\mathbf{x}}}(t, \tau)$ to $\tilde{\mathbf{K}}_{\tilde{\mathbf{x}}}(t, t)$. $\tilde{\mathbf{K}}_{\tilde{\mathbf{x}}}(t, t)$ satisfied the linear matrix differential equation

$$\frac{d\tilde{\mathbf{K}}_{\tilde{\mathbf{x}}}(t, t)}{dt} = \tilde{\mathbf{F}}(t)\tilde{\mathbf{K}}_{\tilde{\mathbf{x}}}(t, t) + \tilde{\mathbf{K}}_{\tilde{\mathbf{x}}}(t, t)\tilde{\mathbf{F}}^+(t) + \tilde{\mathbf{G}}(t)\tilde{\mathbf{Q}}\tilde{\mathbf{G}}^+(t), \tag{B12}$$

where the initial condition $\tilde{\mathbf{K}}_{\tilde{\mathbf{x}}}(T_0, T_0)$ is given as part of the system description; $\tilde{\mathbf{K}}_{\tilde{\mathbf{x}}}(t, \tau)$ is given by

$$\tilde{\mathbf{K}}_{\tilde{\mathbf{x}}}(t, \tau) = \begin{cases}\tilde{\boldsymbol{\theta}}(t, \tau)\tilde{\mathbf{K}}_{\tilde{\mathbf{x}}}(\tau, \tau), & t > \tau, \\ \tilde{\mathbf{K}}_{\tilde{\mathbf{x}}}(t, t)\tilde{\boldsymbol{\theta}}^+(\tau, t), & \tau > t,\end{cases} \tag{B13a}$$

where $\tilde{\theta}(t, \tau)$ is the complex transition matrix associated with $F(t)$. We also note for future reference that

$$\tilde{\mathbf{K}}_{\tilde{x}}(t, \tau) = \tilde{\mathbf{K}}_{\tilde{x}}^{\ +}(\tau, t), \tag{B13b}$$

We can readily show that $\tilde{\mathbf{K}}_{\tilde{x}}(t, t)$ is Hermitian for all t. In order to do this, we simply perform the conjugate transpose operation upon Equation B12. This yields

$$\frac{d\tilde{\mathbf{K}}_{\tilde{x}}^{+}(t, t)}{dt} = \tilde{\mathbf{K}}_{\tilde{x}}^{+}(t, t)\tilde{\mathbf{F}}^{+}(t) + \tilde{\mathbf{F}}(t)\tilde{\mathbf{K}}_{\mathbf{x}}^{+}(t, t) + \tilde{\mathbf{G}}(t)\tilde{\mathbf{Q}}^{+}\tilde{\mathbf{G}}^{+}(t), \tag{B14a}$$

$\tilde{\mathbf{Q}}$ is Hermitian; therefore, $\tilde{\mathbf{K}}_{\tilde{x}}(t, t)$ and $\tilde{\mathbf{K}}_{\tilde{x}}^{+}(t, t)$ satisfy the same linear differential equation. Since $\tilde{\mathbf{K}}_{\tilde{x}}(t, t)$ and $\tilde{\mathbf{K}}_{\tilde{x}}^{+}(t, t)$ have the same initial conditions ($\tilde{\mathbf{K}}_{\tilde{x}}(T_0, T_0)$ is Hermitian by assumption), they must be identical. Consequently, complex covariance matrix $\tilde{\mathbf{K}}_{\tilde{x}}(t, t)$ is Hermitian for all t. We can also show that

$$E[\tilde{\mathbf{x}}(t)\tilde{\mathbf{x}}^T(t)] = \mathbf{0} \tag{B14b}$$

for all t. In order to do this we note that this expectation satisfies the linear differential equation

$$\frac{d}{dt}E[\tilde{\mathbf{x}}(t)\tilde{\mathbf{x}}^T(t)] = \tilde{\mathbf{F}}(t)E[\tilde{\mathbf{x}}(t)\tilde{\mathbf{x}}^T(t)] + E[\tilde{\mathbf{x}}(t)\tilde{\mathbf{x}}^T(t)]\tilde{\mathbf{F}}^T(t)$$
$$+ \tilde{\mathbf{G}}(t)\tilde{\mathbf{Q}}^1\tilde{\mathbf{G}}^T(t). \tag{B15}$$

Since $\tilde{\mathbf{Q}}^1$ equals zero (Equation B8), the forcing term in this equation is zero. In addition, the homogeneous solution is zero for all t (Equation B11). This proves the assertion.

By using the above, we may prove that $E[\tilde{\mathbf{x}}(t)\tilde{\mathbf{x}}^T(\tau)]$ equals zero for all t and τ. To do this we note that

$$E[\tilde{\mathbf{x}}(t)\tilde{\mathbf{x}}^T(\tau)] = \begin{cases} \tilde{\mathbf{\theta}}(t, \tau)E[\tilde{\mathbf{x}}(\tau)\tilde{\mathbf{x}}^T(\tau)], & t > \tau, \\ E[\tilde{\mathbf{x}}(t)\tilde{\mathbf{x}}^T(t)]\tilde{\mathbf{\theta}}^T(t, \tau), & \tau > t. \end{cases} \tag{B16}$$

Since the expectations on the right side of the above equation are zero, the expectation on the left equals zero for all t and τ. We note here that the assumptions that we have made on the covariance of the initial state vector $\tilde{\mathbf{x}}(T_0)$ are satisfied by the covariance of the state vector $\tilde{\mathbf{x}}(t)$ for all $t \geqq T_0$.

Usually, we are not concerned directly with the state vector of a system. The vector interest is the observed signal, $\mathbf{y}(t)$, which is related to the state vector by Equation B3b. We can simply state the properties of the covariance $\tilde{\mathbf{K}}_{\tilde{y}}(t, \tau)$ since it is related directly to the covariance of the state vector. This relationship is given by

$$\tilde{\mathbf{K}}_{\tilde{\mathbf{y}}}(t, \tau) = \tilde{\mathbf{C}}(t)\tilde{\mathbf{K}}_{\tilde{\mathbf{x}}}(t, \tau)\tilde{\mathbf{C}}^{+}(\tau). \tag{B17}$$

Consequently, it is clear that $\tilde{\mathbf{K}}_{\tilde{\mathbf{y}}}(t, t)$ is Hermitian and that $E[\tilde{\mathbf{y}}(t)\tilde{\mathbf{y}}^{+}(\tau)]$ is zero.

We are now in a position to derive some of the properties regarding the individual components of the observed signal. First, let us prove that the covariances $\mathbf{y}_c(t)$ and $\mathbf{y}_s(t)$ are identical. We have

$$\begin{aligned}
E[\mathbf{y}_c(t)\mathbf{y}_c^T(\tau)] &= E\left\{\left(\frac{\tilde{\mathbf{y}}(t) + \tilde{\mathbf{y}}^*(t)}{2}\right)\left(\frac{\tilde{\mathbf{y}}^T(\tau) + \tilde{\mathbf{y}}^{*T}(\tau)}{2}\right)\right\} \\
&= \tfrac{1}{4}(E[\tilde{\mathbf{y}}(t)\tilde{\mathbf{y}}^T(\tau)] + E[\tilde{\mathbf{y}}(t)\tilde{\mathbf{y}}^{*T}(\tau)] \\
&\quad + E[\tilde{\mathbf{y}}(t)\tilde{\mathbf{y}}^{*T}(\tau)] + E[\tilde{\mathbf{y}}(t)\tilde{\mathbf{y}}^T(\tau)]^*) \\
&= \tfrac{1}{2}\,\mathrm{Re}[\mathbf{K}_{\tilde{\mathbf{y}}}(t, \tau)],
\end{aligned} \tag{B18a}$$

$$\begin{aligned}
E[\mathbf{y}_s(t)\mathbf{y}_s(\tau)] &= E\left\{j\left(\frac{\tilde{\mathbf{y}}(t) - \tilde{\mathbf{y}}^*(t)}{2}\right)j\left(\frac{\tilde{\mathbf{y}}^T(\tau) - \tilde{\mathbf{y}}^{*T}(\tau)}{2}\right)\right\} \\
&= \tfrac{1}{4}(-E[\tilde{\mathbf{y}}(t)\tilde{\mathbf{y}}^T(\tau)] + E[\tilde{\mathbf{y}}(t)\tilde{\mathbf{y}}^{*T}(\tau)] \\
&\quad + E[\tilde{\mathbf{y}}(t)\tilde{\mathbf{y}}^{*T}(\tau)]^* - E[\tilde{\mathbf{y}}(t)\tilde{\mathbf{y}}^T(\tau)]^*) \\
&= \tfrac{1}{2}\,\mathrm{Re}[\mathbf{K}_{\tilde{\mathbf{y}}}(t, \tau)].
\end{aligned} \tag{B18b}$$

Consequently, the covariances of both components are equal to one-half the real part of the complex covariance matrix.

The cross covariance between components may be found in a similar fashion. We have

$$\begin{aligned}
E[\mathbf{y}_c(t)\mathbf{y}_s(\tau)] &= E\left\{\left(\frac{\tilde{\mathbf{y}}(t) + \tilde{\mathbf{y}}^*(t)}{2}\right)j\left(\frac{\tilde{\mathbf{y}}^*(\tau) - \tilde{\mathbf{y}}^{*T}(\tau)}{2}\right)\right\} \\
&= \tfrac{1}{4}(E[\tilde{\mathbf{y}}(t)\tilde{\mathbf{y}}^T(\tau)] - E[\tilde{\mathbf{y}}(t)\tilde{\mathbf{y}}^{*T}(\tau)] \\
&\quad + E[\tilde{\mathbf{y}}(t)\tilde{\mathbf{y}}^{*T}(\tau)]^* - E[\tilde{\mathbf{y}}(t)\tilde{\mathbf{y}}^T(\tau)]^*) \\
&= -\tfrac{1}{2}\,\mathrm{Im}[\tilde{\mathbf{K}}_{\tilde{\mathbf{y}}}(t, \tau)].
\end{aligned} \tag{B19}$$

By using Equations B18 and B19, we have a convenient method for finding the auto and cross covariance of the real and imaginary components of the complex signal $\mathbf{y}(t)$ in terms of the complex covariance function $\mathbf{K}_{\tilde{\mathbf{y}}}(t, \tau)$. This provides the notational convenience of working with just one covariance matrix, yet it allows us to determine the covariances of the individual components. This is a major advantage of our complex notation.

B.2 Stationary Random Processes

We now want to show that the assumptions that we made lead to results that are consistent with those that have been developed for stationary scalar random processes. By choosing $\tilde{\mathbf{P}}_0$ to be the steady-state solution to Equation B12, i.e.,

$$\tilde{\mathbf{P}}_0 = \lim_{t \to \infty} \tilde{\mathbf{K}}_{\mathbf{x}}(t, t), \tag{B20}$$

we can show that the covariance matrix $\tilde{\mathbf{K}}\tilde{\mathbf{y}}(t, \tau)$ is stationary. Equivalently, we could say that under this assumption for the initial state vector $\tilde{\mathbf{x}}(T_0)$, $\tilde{\mathbf{y}}(t)$ is a segment of stationary random processes. For stationary random processes we shall use the notation

$$\tilde{\mathbf{K}}_{\tilde{\mathbf{y}}}(\Delta t) = \tilde{\mathbf{K}}_{\tilde{\mathbf{y}}}(t, t + \Delta t); \tag{B21}$$

i.e., we use only one argument.

Since we have already proved that the real and imaginary components have the same covariance (Equation B18), then Equation B2a is certainly satisfied. Therefore, we need only to prove that the cross covariance is an odd function. In general, we have (Equation B14),

$$\begin{aligned} &\mathrm{Re}[\tilde{\mathbf{K}}_{\tilde{\mathbf{y}}}(\Delta t)] + j\,\mathrm{Im}[\tilde{\mathbf{K}}_{\tilde{\mathbf{y}}}(\Delta t)] \\ &\quad = \mathrm{Re}[\tilde{\mathbf{K}}_{\tilde{\mathbf{y}}}(\Delta t)]^T - j\,\mathrm{Im}[\tilde{\mathbf{K}}_{\tilde{\mathbf{y}}}(\Delta t)]^T. \end{aligned} \tag{B22a}$$

By equating the imaginary parts and substituting Equation B19, we obtain

$$\tilde{\mathbf{K}}_{\mathbf{y}_c\mathbf{y}_s}(\Delta t) = -\tilde{\mathbf{K}}^T_{\mathbf{y}_c\mathbf{y}_s}(\Delta t) = -\tilde{\mathbf{K}}_{\mathbf{y}_s\mathbf{y}_c}(-\Delta t), \tag{B22b}$$

which for the scalar case is consistent with Equation B2b. These conditions are sufficient to show that $\mathbf{y}(t)$ has a real, positive spectrum in the scalar case.

B.3 Summary

In this section we introduced the idea of generating a complex random process by exciting a linear system having a complex state variable description with a complex white noise. We then showed how we could describe the second-order statistics of this process in terms of a complex covariance function and then we discussed how we could determine this function from the state variable description of the system. The only assumptions that we were required to make were on $\tilde{\mathbf{u}}(t)$ and $\tilde{\mathbf{x}}(T_0)$. Our results were independent of the form of the coefficient matrices $\mathbf{F}(t)$, $\mathbf{G}(t)$, and $\mathbf{C}(t)$. Our methods were exactly parallel to those for real state variables, and our results were similar to those that have been developed for describing narrow-band processes by complex notation. We shall now consider two examples to illustrate the type of random processes that we may generate.

Example B.1

The first example we consider is the first-order (scalar) case. The equations that describe this system are

$$\frac{d\tilde{x}(t)}{dt} = -\tilde{k}\tilde{x}(t) + \tilde{u}(t) \qquad \text{(state equation)}, \tag{B23}$$

and

$$\tilde{y}(t) = \tilde{x}(t) \qquad \text{(observation equation)} \tag{B24}$$

The assumptions regarding $\tilde{\mathbf{u}}(t)$ and $\tilde{\mathbf{x}}(T_0)$ are

$$E[\tilde{u}(t)\tilde{u}^*(\tau)] = 2\,\mathrm{Re}[\tilde{k}]P\delta(t-\tau) \tag{B25}$$

and

$$E[|\tilde{x}(T_0)|^2] = P_0. \tag{B26}$$

Since we have a scalar process, both P and P_0 must be real. In addition, we have again assumed zero means.

First, we shall find $\tilde{K}_x(t, t)$. The differential equation, Equation B12, which it satisfies is

$$\frac{d\tilde{K}_{\tilde{x}}(t, t)}{dt} = -\tilde{k}K_{\tilde{x}}(t, t) - \tilde{k}^* K_{\tilde{x}}(t, t) + 2\,\mathrm{Re}[\tilde{k}]P$$
$$= 2\mathrm{Re}[\tilde{k}]\tilde{K}_{\tilde{x}}(t,t) + 2\mathrm{Re}[\tilde{k}]P. \tag{B27}$$

The solution to this equation is

$$\tilde{K}_{\tilde{x}}(t, t) = P(P - P_0)\mathrm{e}^{-2Re[\tilde{k}](t-T_0)}, \qquad t > T_0. \tag{B28}$$

In order to find $\tilde{K}_x(t, \tau)$, we need to find $\theta(t, \tau)$, the transition matrix for this system. This is easily found to be

$$\tilde{\theta}(t, \tau) = e^{-\tilde{k}(t-\tau)}. \tag{B29}$$

By substituting Equations B28 and B29 into Equation B13, we find $\tilde{K}_{\tilde{x}}(t, \tau)$, which is also $\tilde{K}_y(t, \tau)$ for this particular example. Furthermore, we find the auto and cross covariance of the individual components by applying Equations B18 and B19, respectively.

Let us now consider the stationary case in more detail. In this case

$$P_0 = P. \tag{B30}$$

If we perform the indicated substitutions, we obtain

$$\tilde{K}_{\tilde{x}}(\Delta t) = \begin{cases} Pe^{-\tilde{k}\,\Delta t}, & \Delta t \geqq 0, \\ Pe^{\tilde{k}^*\,\Delta t}, & \Delta t \leqq 0. \end{cases} \tag{B31}$$

This may be written as

$$\tilde{K}_{\tilde{x}}(\Delta t) = Pe^{-\mathrm{Re}[\tilde{k}]|\Delta t|}e^{-j\,\mathrm{Im}[\tilde{k}]\,\Delta t}. \tag{B32}$$

By applying Equations B18 and B19 we find

$$K_{x_c x_c}(\Delta t) = K_{x_s x_s}(\Delta t) = \frac{P}{2}\,e^{-\mathrm{Re}[\tilde{k}]|\Delta t|}\cos(\mathrm{Im}[\tilde{k}]\,\Delta t), \tag{B33a}$$

$$K_{x_c x_s}(\Delta t) = \frac{P}{2}\,e^{-\mathrm{Re}[\tilde{k}]|\Delta t|}\,\sin(\mathrm{Im}[k]\,\Delta t). \tag{B33b}$$

The spectrum of the complex process follows easily as

$$S_{\tilde{x}}(\omega) = \frac{2\mathrm{Re}[k]P}{(\omega - \mathrm{Im}[k])^2 + \mathrm{Re}[k]^2}. \tag{B34}$$

The spectra and cross spectra of $x_c(t)$ and $x_s(t)$ may easily be found in terms of the even and odd parts of $S_{\tilde{x}}(\omega)$.

From Equation B34, we see that, in the stationary case, the net effect of allowing a complex gain is that the spectrum has a frequency shift equal to the imaginary part of the gain. In a narrow-band interpretation this would correspond to a mean Doppler shift about the carrier. In general, we would not expect such a simple interpretation of the effect a complex state representation. This suggests that we should consider a second example where we have a second-order system and two feedback gains.

Example B.2

In this example we want to analyze a second-order system. In the steady state, it corresponds to a system with two poles. Note that the pole locations $-k_1$ and $-k_2$ need not be complex conjugates of one another. Again, we shall analyze the stationary case; the analysis for the nonstationary case is straightforward, but is not especially informative. The state and observation equations for this system are

$$\frac{d}{dt}\begin{bmatrix}\tilde{x}_1(t)\\ \tilde{x}_2(t)\end{bmatrix} = \begin{bmatrix} 0 & 1 \\ -\tilde{k}_1\tilde{k}_2 & -(\tilde{k}_1 + \tilde{k}_2)\end{bmatrix}\begin{bmatrix}\tilde{x}_1(t)\\ \tilde{k}_2(t)\end{bmatrix} + \begin{bmatrix}0\\1\end{bmatrix}\tilde{u}(t) \tag{B35}$$

and

$$\tilde{y}(t) = [1 : 0]\begin{bmatrix}\tilde{x}_1(t)\\ \tilde{x}_2(t)\end{bmatrix}. \tag{B36}$$

The covariance of the driving noise is

$$E[\tilde{u}(t)\tilde{u}^*(\tau)] = 2P\Bigg(\text{Re}[\tilde{k}_1\tilde{k}_2]\text{Re}[\tilde{k}_1 + \tilde{k}_2] + \frac{\text{Im}[\tilde{k}_1\tilde{k}_2]}{\text{Re}[\tilde{k}_1 + \tilde{k}_2]}[\text{Im}(\tilde{k}_1 - \tilde{k}_2)\text{Re}(\tilde{k}_1 + \tilde{k}_2) - \text{Im}(\tilde{k}_1\tilde{k}_2)]\Bigg)\delta(t-\tau). \tag{B37}$$

From the steady state solution to Equation B12, we have

$$\tilde{\mathbf{P}}_0 = P\begin{bmatrix} 1 & \dfrac{j\,\text{Im}[\tilde{k}_1\tilde{k}_2]}{\text{Re}[\tilde{k}_1 + \tilde{k}_2]} \\ \dfrac{-j\,\text{Im}[\tilde{k}_1\tilde{k}_2]}{\text{Re}[\tilde{k}_1 + \tilde{k}_2]} & \dfrac{(\text{Re}(\tilde{k}_1\tilde{k}_2)\text{Re}[\tilde{k}_1 + \tilde{k}_2] + \text{Im}(\tilde{k}_1\tilde{k}_2)\text{Im}(\tilde{k}_1 + \tilde{k}_2))}{\text{Re}[\tilde{k}_1 + \tilde{k}_2]} \end{bmatrix} \tag{B38}$$

Therefore, we have a stationary process $\tilde{y}(t)$ with power P.

In order to find the covariance matrix we need to find the transition matrix for the system. We do this by using matrix Laplace transform techniques. We find

$$\tilde{\boldsymbol{\theta}}(t,\tau) = \frac{1}{\tilde{k}_2 - \tilde{k}_1}\begin{bmatrix} [\tilde{k}_2 e^{-\tilde{k}_1(t-\tau)} - k_1 e^{-\tilde{k}_2(t-\tau)}] & e^{-\tilde{k}_1(t-\tau)} - e^{-\tilde{k}_2(t-\tau)} \\ -(\tilde{k}_1\tilde{k}_2)[e^{-\tilde{k}_1(t-\tau)} - e^{-\tilde{k}_2(t-\tau)}] & -[\tilde{k}_1 e^{-(\tilde{k}_1(t-\tau)} - \tilde{k}_2 e^{-\tilde{k}_2(t-\tau)}] \end{bmatrix} \tag{B39}$$

If we substitute Equations B38 and B39 in Equation B13 and then use Equation B17, we obtain

$$\tilde{K}_{\tilde{k}}(\Delta t) = \begin{cases} \dfrac{P}{(\tilde{k}_2 - \tilde{k}_1)^*}(\tilde{k}_2^* e^{k_1^*\Delta_1} - \tilde{k}_1^* e^{k_2^*\Delta t}) + \dfrac{j\,\text{Im}[\tilde{k}_1\tilde{k}_2]}{\text{Re}[\tilde{k}_1 + \tilde{k}_2]}(e^{k_1^*\Delta t} - e^{k_2^*\Delta t}) & \Delta t < 0, \\ \dfrac{P}{(\tilde{k}_2 - \tilde{k}_1)}(\tilde{k}_2 e^{-\tilde{k}_1\Delta_t} - \tilde{k}_1 e^{-\tilde{k}_2\Delta t}) - \dfrac{j\,\text{Im}[\tilde{k}_1\tilde{k}_2]}{\text{Re}[\tilde{k}_1 + \tilde{k}_2]}(e^{-\tilde{k}_1\Delta_t} - e^{-\tilde{k}_2\Delta t}) & \Delta t > 0. \end{cases} \tag{B40}$$

We now want to determine the spectrum $S_y(\omega)$ from Equation B40. Let us define two coefficients for convenience:

$$\tilde{A}_1 = \frac{\tilde{k}_2 + \tilde{k}_1^*}{(\tilde{k}_2 - \tilde{k}_1)} \cdot \frac{\text{Re}[\tilde{k}_2]}{\text{Re}[\tilde{k}_1 + \tilde{k}_2]}, \tag{B41}$$

$$\tilde{A}_2 = -\frac{\tilde{k}_2^* + \tilde{k}_1}{(\tilde{k}_2 - \tilde{k}_1)} \cdot \frac{\text{Re}[\tilde{k}_1]}{\text{Re}[\tilde{k}_1 + \tilde{k}_2]}. \tag{B42}$$

After a fair amount of manipulation, we can compute $S_y(\omega)$.

$$S_y(\omega) = 2P\left(\frac{\text{Re}[\tilde{A}_1]\text{Re}[\tilde{k}_1] + \text{Im}[\tilde{A}_1](\omega + \text{Im}[\tilde{k}_1])}{\text{Re}^2[\tilde{k}_1] + (\omega + \text{Im}[\tilde{k}_1])^2}\right.$$
$$\left. + \frac{\text{Re}[\tilde{A}_2]\text{Re}[\tilde{k}_2] + \text{Im}[\tilde{A}_2](\omega + \text{Im}[\tilde{k}_2])}{\text{Re}^2[\tilde{k}_2] + (\omega + \text{Im}[\tilde{k}_2])^2}\right). \tag{B43}$$

We have plotted this function for various values of $\tilde{k}_1$ and $\tilde{k}_2$ in Figures B.1 through B.4. The values of $\tilde{k}_1$ and $\tilde{k}_2$ for a particular figure are illustrated on the figures by the pole location they produce; i.e., the system has a pole at $-\tilde{k}_1$ and $-\tilde{k}_2$.

In Figures B.1 and B.2 we illustrate that by simply choosing the poles as complex conjugates we can produce a spectrum which is either very flat near $\omega = 0$ or is peaked with two symmetric lobes. If one wanted to use real-state variables to generate this spectrum, a fourth-order system would be required. In this case, the complex notation has significantly reduced the computation required.

Figure B.3 illustrates an interesting observation about mean Doppler shifts. Let us draw the pole-zero locations for the complex system. If there exists a line $\omega = \omega_c$ about which the pole-zero pattern is symmetric, then in the stationary case the complex notation effectively produces a spectrum which is symmetric about $\omega = \omega_c$. For example, in Figure B.3, the pole pattern is symmetric about $\omega = 1/2$. We see that the spectrum is symmetric about $\omega = -1/2$ also. Consequently, we can use the complex notation to introduce mean Doppler shifts of narrow-band processes.

Figure B.4 illustrates that we can obtain spectra that are not symmetric about any axis. This is a relatively important case. In dealing with narrow-band processes, if one can find a frequency about which the spectrum is symmetric, then the component processes $y_c(t)$ and $y_s(t)$ are uncorrelated. However, if there is no axis of symmetry (or if the choice of carrier is not at our disposal), then the components are definitely correlated. This example shows that we can model narrow-band processes with nonsymmetric spectra very conveniently with our complex state variable notation.

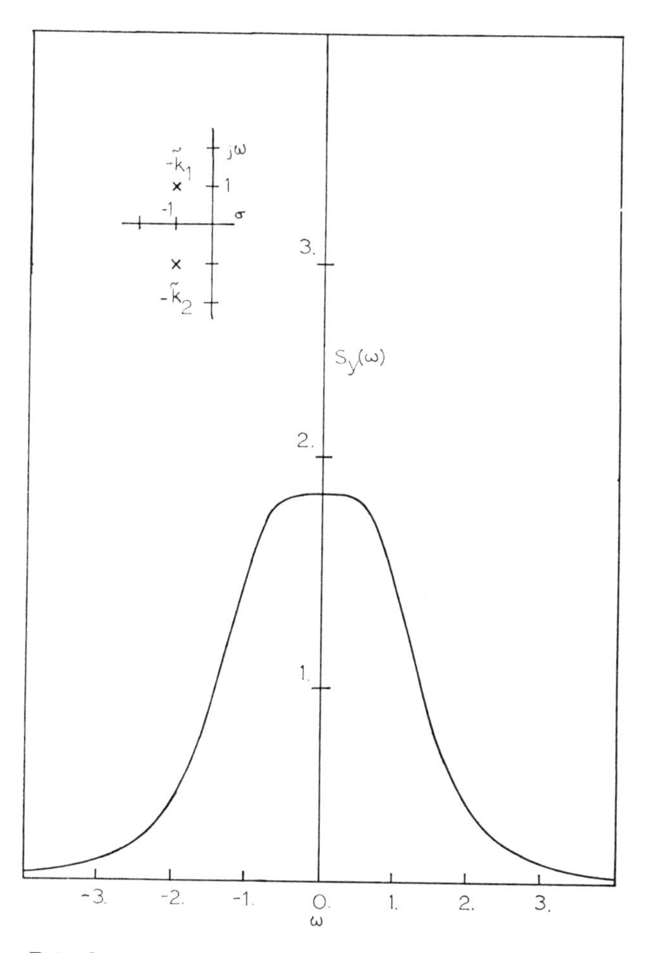

Figure B.1 Spectrum for a second-order complex process when poles are closely spaced.

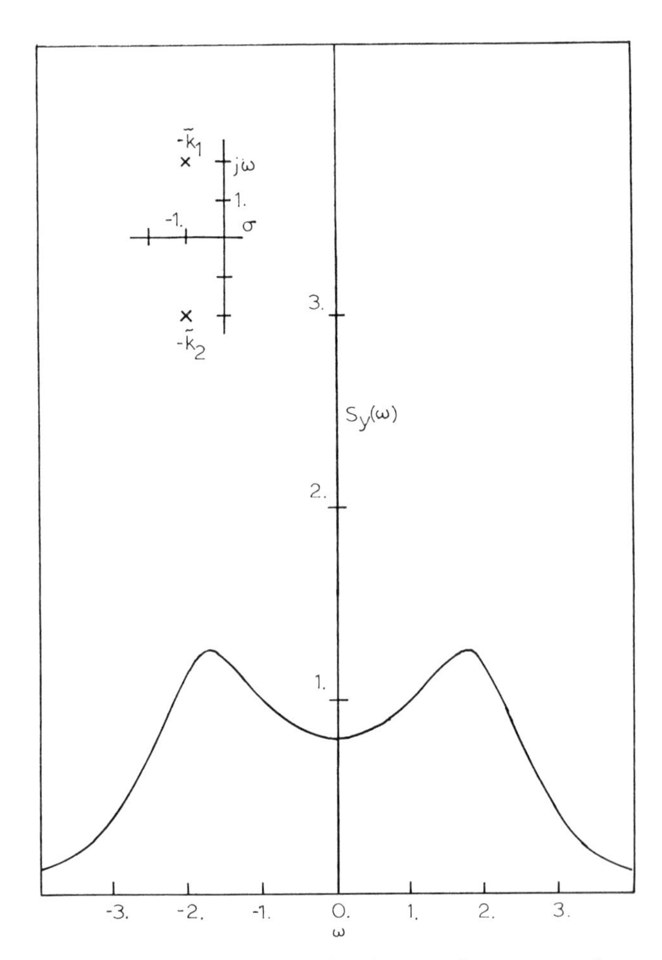

Figure B.2 Spectrum for a second-order complex process when poles are widely spaced.

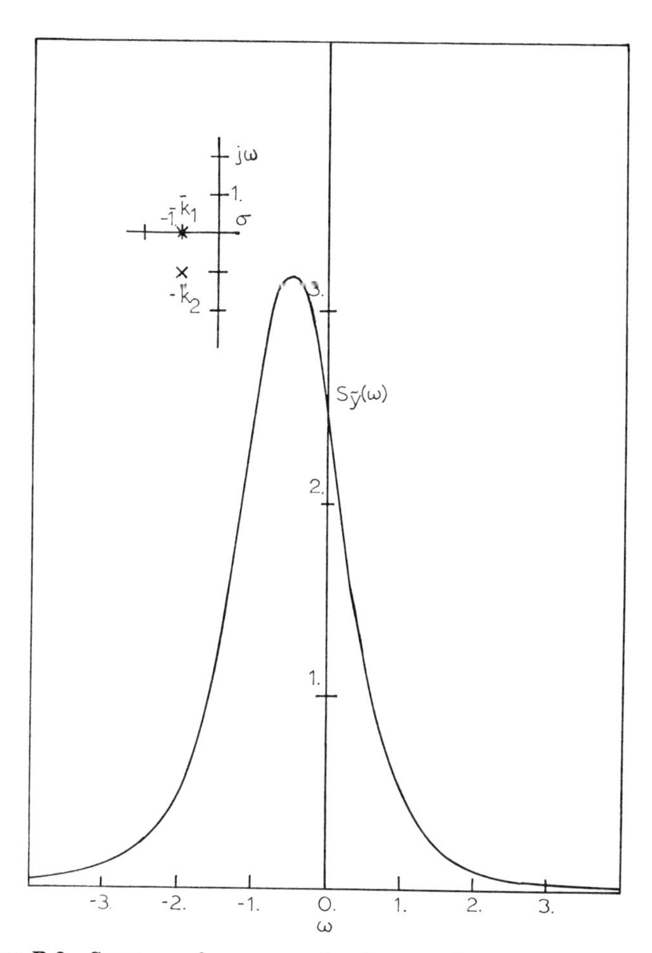

Figure B.3 Spectrum for a second-order complex process with a mean Doppler frequency shift.

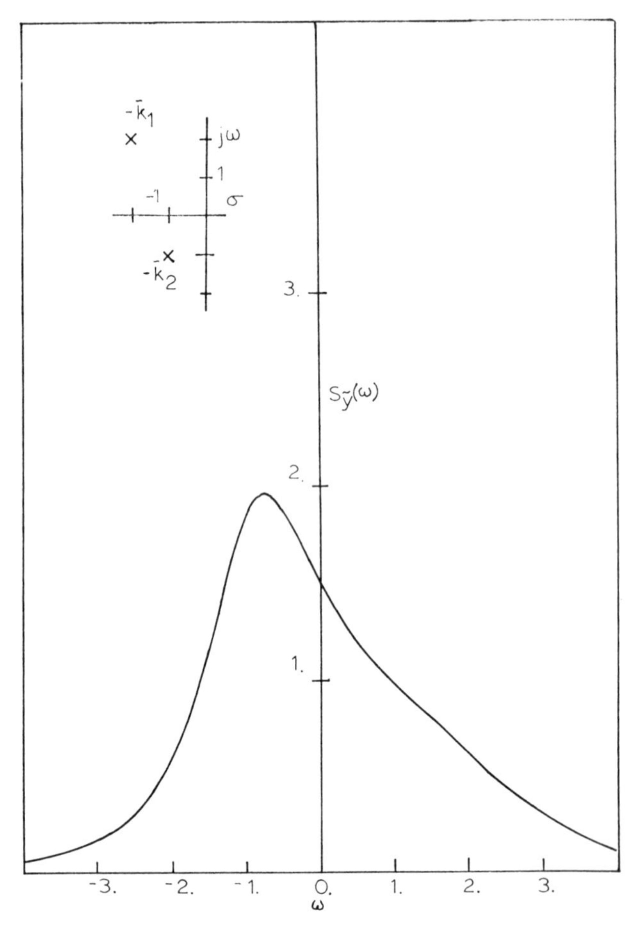

Figure B.4 Spectrum for a second-order complex process with non-symmetrical pole locations.

Bibliography

1. B. D. O. Anderson, J. B. Moore, and S. G. Loo, "Time Varying Spectral Factorization, 1: Construction of Spectral Factors," Technical Report EE-6701, University of New Castle, New South Wales, Australia, June 1967.
2. M. Athans, "The Status of Optimal Control Theory and Applications for Deterministic Systems," 1966 IEEE International Convention Record, March 1966.
3. M. Athans and P. L. Falb, *Optimal Control*, McGraw-Hill Book Company, Inc., New York, 1966.
4. M. Athans and F. C. Schweppe, "Gradient Matrices and Matrix Calculations," Technical Note 1965-53, M.I.T. Lincoln Laboratory, Cambridge, Mass., 1965.
5. M. Athans and F. C. Schweppe, "Optimal Waveform Design via Control Theoretic Concepts," *Inform. Control*, **10** (4), 335–377 (1967).
6. A. Baggeroer, "Maximum Aposteriori Interval Estimation," 1966 WESCON Convention Record, Paper 7-3, August 1966.
7. A. Baggeroer, *A State Variable Technique for Solving Fredholm Integral Equations*, Technical Report 459, Research Laboratory of Electronics, Massachusetts Institute of Technology, Cambridge, Mass., 1968.
8. A. B. Baggeroer, L. D. Collins, and H. L. Van Trees, "Complex State Variables, Theory, and Applications," 1968 WESCON Proceedings, Session 6-3.
9. A. B. Baggeroer, "State Variables, the Fredholm Theory, and Optimal Communication," Sc.D. Thesis, Department of Electrical Engineering, Massachusetts Institute of Technology, Cambridge, Mass., 1968.

10. A. B. Baggeroer, "On the Performance of the Optimal Smoother," Research Laboratory of Electronics, Quarterly Progress Report, Massachusetts Institute of Technology, Cambridge, Mass., 15 June, 1968.
11. A. B. Baggeroer, "Optimal Signal Design for Additive Colored Noise Channels via State Variables," 1968 WESCON Proceedings, Session 6-3.
12. A. V. Balakrishnan, "Signal Design for a Class of Clutter Channels," *IEEE Trans. Inform. Theory*, **IT-14** (1), 170–173 (1968).
13. R. Bellman, *Invariant Imbedding*, Academic Press Inc., New York, 1964.
14. R. Bellman, R. Kalaba, and R. Sridhar, "Sensitivity Analysis and Invariant Imbedding," Research Memorandum No. 4039-PR, Rand Corporation, March 1964.
15. A. E. Bryson and M. Frazier, "Smoothing for Linear and Nonlinear Dynamic Systems," Proceedings of Optimum Synthesis Conference, Wright Patterson Air Force Base, September 1962.
16. J. Capon, "Asymptotic Eigenfunctions and Eigenvalues of a Homogeneous Integral Equation," *IRE Trans. Inform. Theory*, **IT-8** (1), 2–4 (1962).
17. L. D. Collins, "A Simplified Derivation of the Differential Equation Solution to Fredholm Integral Equations of the Second Kind," Internal Memo No. IM-LDC-19, Detection and Estimation Theory Group, Research Laboratory of Electronics, M.I.T., Cambridge, Mass., January 1967.
18. L. D. Collins, "Realizable Whitening Filters and State Variable Realizations," *Proc. IEEE*, **68** (1), 100–101 (1968).
19. L. D. Collins, "Evaluation of the Fredholm Determinant for State-Variable Covariance Functions," *Proc. IEEE*, **56** (3), 350–351 (1968).
20. L. D. Collins, "Asymptotic Approximation to the Error Probability for Detecting Gaussian Signals," Sc.D. Thesis, Department of Electrical Engineering, Massachusetts Institute of Technology, Cambridge, Mass., June 1968.
21. R. Courant and D. Hilbert, *Methods of Mathematical Physics, Vol. I*, Interscience Publishers, New York, 1953.
22. W. Davenport and W. Root, *Random Signals and Noise*, McGraw-Hill Book Company, Inc., New York, 1958.
23. D. F. Delong, Jr., and E. M. Hofstetter, "On the Design of Optimum Radar Waveforms for Clutter Rejection," *IEEE Trans. Inform. Theory*, **IT-13**, 454–463 (1967).
24. P. M. DeRusso, R. J. Roy, and C. M. Close, *State Variables for Engineers*, John Wiley & Sons, Inc., New York, 1956.
25. D. M. Detchmendy and Sridar, "Sequential Estimation of States and Parameters in Noisy Dynamical Systems," *ASME J. Basic Eng.*, **88**, 362–368 (1966).
26. W. Everling, "On the Evaluation of e^{AT} by Power Series," *Proc. IEEE* (*Letters*), **55**, 413 (1967).
27. V. N. Faddeeva, *Computational Methods of Linear Algebra*, Dover Publications, Inc., New York, 1959.

28. D. C. Fraser, "A New Technique for the Optimal Smoothing of Data," Sc.D. Thesis, Department of Aeronautical Engineering, Massachusetts Institute of Technology, Cambridge, Mass., January, 1967.
29. C. W. Helstrom, *Statistical Theory of Signal Detection*, Pergamon Press, Inc., London, 1960.
30. C. W. Helstrom, "Solution of the Detection Integral Equation for Stationary Filtered White Noise," *IEEE Trans. Inform. Theory*, **IT-11** (3), 335–339 (1965).
31. T. E. Hinrichs, "Iterative Methods for Evaluation of the Transition Matrix," Sc.B. Thesis, Department of Electrical Engineering, Massachusetts Institute of Technology, Cambridge, Mass., February 1969.
32. J. M. Holtzman, "Signal-to-Noise Ratio Maximization Using Pontryagin's Maximum Principle," *Bell System Tech. J.*, **45**, 473–489 (March 1966).
33. R. Y. Huang and R. A. Johnson, "Information Transmission with Time Continuous Processes," *IEEE Trans. Inform. Theory*, **IT-9** (2), 84–95 (1963).
34. T. Kailath, "Some Integral Equations with Non-Rational Kernels," *IEEE Trans. Inform. Theory*, **IT-12** (4), 442–447 (1966).
35. R. E. Kalman and R. Bucy, "New Results in Linear Filtering and Prediction Theory," *ASME J. Basic Eng.*, **83**, 95–108 (1961).
36. E. J. Kelley and W. L. Root, "A Representation for Vector Valued Random Processes," Group Report 55-21 (revised), M.I.T. Lincoln Laboratory, Cambridge, Mass., 22 April, 1960.
37. R. S. Kennedy, *Communication over Fading Dispersive Channels*, John Wiley & Sons, Inc., New York, 1969.
38. T. G. Kincaid, "On Optimum Waveforms for Correlation Detection in the Sonar Environment: Reverberation Limited Conditions," *J. Acoust. Soc. Am.*, **46** (3), 787–796 (1968).
39. T. G. Kincaid, "Optimum Waveforms for Correlation Detection in the Sonar Environment: Noise Limited Conditions," *J. Acoust. Soc. Am.*, **43**, (2), 258–268 (1968).
40. H. J. Kushner, "On the Dynamical Equations of Conditional Probability Density Functions with Applications to Optimal Stochastic Control Theory," *J. Math. Anal. Appl.*, **8**, 332–344 (1964).
41. H. J. Kushner and F. C. Schweppe, "A Maximum Principle for Stochastic Control Systems," *J. Math. Anal. Appl.*, **8**, 287–302 (1964).
42. J. H. Laning and R. H. Battin, *Random Processes in Automatic Control*, McGraw-Hill Book Company, Inc., New York, 1956.
43. M. L. Liou, "Evaluation of the Transition Matrix," *Proc. IEEE*, **55** (2), 228–229 (1967).
44. J. S. Meditch, "On Optimal Linear Smoothing Theory," Technical Report 67-105, Information Processing and Control System Laboratory, Northwestern University, March 1967.
45. D. Middleton, *Introduction to Statistical Communication Theory*, McGraw-Hill Book Company, Inc., New York, 1960.

46. A. Papoulis, *Probability, Random Variables, and Stochastic Processes,* McGraw-Hill Book Company, Inc., New York, 1965.
47. L. S. Pontryagin, V. G. Boltyanaki, R. Gamkrelidze, and E. Mishenko, *The Mathematical Theory of Optimal Processes*, Interscience Publishers, New York, 1962.
48. A. Post, "Optimal Signal Design for a Correlation Detector," Sc.M. Thesis, Department of Electrical Engineering, Massachusetts Institute of Technology, Cambridge, Mass., August 1968.
49. H. Rauch, F. Tung, and C. Striebel, "Maximum Likelihood Estimates of Linear Dynamic Systems," *AIAA J.*, **3** (8), 1445–1450 (1966).
50. W. L. Root, "Stability in Signal Detection Problems," Stochastic Processes in Mathematical Physics and Engineering, Proceedings Symposium of Applied Mathematics, No. 16, 1964.
51. W. D. Rummler, "Clutter Suppression by Complex Weighting of Coherent Pulse Trains," *IEEE Trans. Aerospace Electron.*, **2** (6), 689 (1966).
52. F. Schweppe, "Evaluation of Likelihood Functions for Gaussian Signals," *IEEE Trans. Inform. Theory*, **IT-11** (1), 61–70 (1965).
53. F. C. Schweppe and D. Gray, "Radar Signal Design Subject to Simultaneous Peak and Average Power Constraints," *IEEE Trans. Inform. Theory*, **12** (1), 13–26 (1966).
54. W. M. Siebert, *Course Notes for* 6.05, Chapter 2, Department of Electrical Engineering, Massachusetts Institute of Technology, Cambridge, Mass.
55. A. J. F. Siegert, "Passage of Stationary Processes through Linear and Nonlinear Devices," *IRE Trans. Inform. Theory*, **PGIT-3**, 4–25 (1954).
56. A. J. F. Siegert, "A Systematic Approach to a Class of Problems in the Theory of Noise and Other Random Phenomena—Part III, Examples," *IRE Trans. Inform. Theory*, **IT-4** (1), 3–14 (1958).
57. A. J. F. Siegert, "A Systematic Approach to a Class of Problems in the Theory of Noise and Other Random Phenomena—Part II, Examples," *IRE Trans. Inform. Theory*, **IT-3** (1), 38–43 (1967).
58. D. Slepian, "Estimation of Signal Parameters in the Presence of Noise," *IRE Trans. Inform. Theory*, **PGIT-3** (68), 68–89 (1954).
59. D. Slepian, H. J. Landau, and H. O. Pollak, "Prolate Spheroidal Wave Functions, Fourier Analysis and Uncertainty: Parts I and II," *Bell System Tech. J.*, **40**, 43–65 (1961); **41**, 65–84 (1961).
60. L. J. Spafford, "Optimum Radar Signal Processing in Clutter," *IEEE Trans. Inform. Theory*, **14** (5), 734 (1968).
61. D. L. Snyder, *The State-Variable Approach to Continuous Estimation with Applications to Analog Communication Theory*, M.I.T. Press, Cambridge, Mass., 1969.
62. J. A. Stratton, P. M. Morse, L. D. Chu, J. D. C. Little, and F. J. Corbato, *Spheroidal Wave Functions*, M.I.T. Press, Cambridge, Mass., 1956.
63. J. S. Thompson and E. L. Titlebaum, "Design of Optimal Radar Waveforms for Clutter Rejection Using the Maximum Principle," *IEEE Trans. Aerospace Electron.*, **3** (6), 581–589 (1967).

64. E. C. Titchmarsh, *Theory of Functions*, Oxford University Press, New York, 1939.
65. D. Tufts and Schnidman, "Optimal Waveform Subject to Both Energy and Peak Value Constraint," *Proc. IEEE*, **52**, 1002–1007 (1965).
66. H. L. Van Trees, "Optimum Signal Design and Processing for Reverbaration-Limited Environments," *IEEE Trans. Military Electron.* **MIL-9** (3, 4), 221–229 (1965).
67. H. L. Van Trees, *Detection, Estimation, and Modulation Theory, Part I*, John Wiley & Sons, Inc., New York, 1968.
68. H. L. Van Trees, *Detection, Estimation, and Modulation Theory, Part II*, John Wiley & Sons, Inc. New York (to appear).
69. H. L. Van Trees (private communication).
70. A. J. Viterbi, "Performance of an M-Ary Orthogonal Communication System Using Stationary Stochastic Signals," *IEEE Trans Inform. Theory*, **IT-13** (3), 414–422 (1967).
71. C. L. Weber, *Elements of Detection and Signal Design*, McGraw-Hill Book Company, Inc., New York, 1968.
72. P. A. Wintz and A. J. Kurtenback, "Waveform Error Control in PCM Telemetry," *IEEE Trans. Inform. Theory*, **IT-14**, (5), 650–661 (1968).
73. D. Youla, "The Solution of a Homogeneous Wiener-Hopf Integral Equation Occurring in the Expansion of Second-Order Stationary Random Functions," *IRE Trans. Inform. Theory*, **IT-3** (4), 187–193 (1957).
74. L. A. Zadeh and J. R. Ragazzini, "An Extension of Wiener's Theory of Prediction," *J. Appl. Phys.*, **21** (7), 645–655 (1950).
75. L. A. Zadeh and K. S. Miller, "Solution to an Integral Equation Occurring in the Theories of Prediction and Detection," *IRE Trans. Inform. Theory*, **IT-2** (3), 72–76 (1956).
76. L. A. Zadeh and C. Desoer, *Linear System Theory*, McGraw-Hill Book Company, Inc., New York, 1966.

Name Index

Anderson, B.D.O., 17
Athans, M., 24, 33, 84, 116, 118, 119

Baggeroer, A. B., 32, 96, 123, 134, 145, 172
Balakrishnan, A., 120
Battin, R., 23
Bellman, R., 164
Beltyanski, V. G., 84
Bryson, A. E., 56, 63, 122, 127, 134, 161
Bucy, R., 2, 4, 65–66, 122, 169

Capon, J., 41
Chu, L. J., 44
Close, C. M., 8
Collins, L. D., 24, 32, 62, 128, 172
Corbato, F. J., 44
Courant, R., 31

Davenport, W., 3, 10, 22, 59
Delong, D. F., 120
DeRusso, P. M., 8
Desoer, C., 8, 17
Detchmendy, D. M., 164
Delong, D., 120

Everling, W., 171

Faddeeva, V. N., 29, 43, 171
Falb, P., 8, 84
Frazier, M., 56, 63, 122, 127, 134, 161

Gamkrelidze, R., 84
Gray, D., 119

Helstrom, C. W., 23
Hilbert, D., 31
Hinrichs, T. E., 171
Hofstetter, E., 120
Holtzman, J. M., 119
Huang, R. Y., 41

Johnson, R. A., 41

Kailath, T., 38
Kalman, R. E., 2, 4, 65–66, 122, 169
Kelley, E. J., 23
Kennedy, R. S., 120
Kincaid, T. G., 120
Kurtenback, A. J., 31
Kushner, H. J., 2, 29, 156

Landau, H. J., 44
Laning, J. H., 23
Liou, M. L., 29, 43, 170

Meditch, J. S., 168
Middleton, D., 23, 24
Miller, K. S., 23
Mishenko, E., 84
Morse, P. M., 44

Papoulis, A., 131
Pollak, H. O., 44

Pontryagin, L. S., 84
Post, A. E., 48

Ragazzini, J. R., 23
Rauch, H., 56, 63, 122, 132, 134
Root, W. L., 3, 10, 22, 59
Roy, R. J., 8
Rummler, W. D., 120

Schnidman, 199
Schweppe, F. C., 2, 24, 33, 119
Siebert, W. M., 149
Siegert, A. J. F., 34, 78
Slepian, D., 44
Snyder D. L., 2, 9, 156, 161, 167, 169
Spafford, L. J., 120
Sridhar, R., 164
Stratton, J. A., 44
Striebel, C., 56, 63, 122, 132, 134

Titchmarsh, E. C., 32
Titlebaum, E. L., 119
Thompson, J. S., 119
Tufts, D., 119
Tung, F., 56, 63, 122, 132, 134

Van Trees, H. L., 3, 10, 23, 24, 28, 31, 60, 63, 83, 84, 92, 93, 96, 118, 120, 121, 155, 162, 172
Viterbi, A. J., 24

Wiener, N., 2
Weber, C., 120
Wintz, P. A., 31

Youla, D. C., 43

Zadeeh, L. A., 8, 17, 23

Subject Index

Adjoint systems, 20, 130
Ambiguity function, 120
Asymptotic expressions
  eigenvalues of high index, 41
  Fredholm determinant, 34
  large time intervals, 77
  signal design, 85, 91, 94–95

Bandlimited processes, ideal, 44, 52–53
Bandpass process representation, 10, 60, 116, 172–186
  advantage of complex notation for, 177
Bandwidth, 74, 76. *See also* Constraints for signal design
Boundary conditions
  for covariance operator, 21
  for homogeneous integral equations, 27–28
  for inhomogeneous integral equation, 62, 64, 66–67
  for linear smoother, 123
  for minimum principle, 89
  for nonlinear smoother, 168
  for signal design, 91
Butterworth processes
  first-order, 15, 36–37, 72–75, 102–109, 136–140, 153-155, 179–181
  higher-order, 43–53

Calculus of variations, for signal design, 84, 118
Carrier frequency. *See* Bandpass process representation
Characteristic polynomials, 28
Clutter. *See* Detection in clutter
Coefficient matrices. *See* Stability of solutions
Colored noise. *See* Detection in colored noise
Combining signal design solutions, 99
Complex state variables. *See* Bandpass process representation
Continuous time issues, 9, 132
Control functions, 85, 88
Constant parameter systems, 14, 133. *See also* Transition matrices
Constraints for signal design
  amplitude, 81, 83, 87, 118
  bandwidth, 81, 83, 87, 118
  hard amplitude, 119
  hard bandwidth, 114–117
Coupling in nonlinear estimators, 169
Covariance matrix
  process
    complex, 173
    operator, 18–22
    properties, 10–18
  error
    filter with delay, 149–155
    realizable filter, 32, 66, 121, 128, 131–132

Covariance matrix (*continued*)
smoother, 128–134

Degradation. *See* Detection, performance
Delay filtering. *See* Filtering, delay
Detection
in clutter, 60, 82, 119
in colored noise, 3, 5, 57–60, 82
performance, 31, 55, 62–63, 82, 100
bounds, 31, 35
comparison for signal design, 109–110, 113
receivers, 3, 5, 57–60, 82
d_g^2-B^2 (degradation-bandwidth) plane, 101–103, 110
Differential equations. *See also* Covariance matrix
filter with delay, 146
homogeneous integral equation, 27–28
inhomogeneous integral equations, 62
linear smoother, 123
linear realizable filter, 127
nonlinear realizable filter, 168
nonlinear smoother, 163
signal design, 90–91
d^2 (*d*-squared). *See* Detection, performance
Doppler spread channels, 60, 82, 96, 116–120. *See also* Detection in clutter
Dynamical system description. *See also* Differential equations; Boundary conditions
process generation, 8–11

Eigenfunctions. *See also* Integral equations, homogeneous
expansion for nonlinear filtering, 162
Eigenvalue. *See also* Integral equations, homogeneous
bounds on largest, 28
equation, 26
smallest re signal design, 81, 100, 120
Envelope, complex, 173
Error probabilities. *See* Detection, performance
Errors re smoother error covariance, 134
Estimators. *See* Filtering; Smoothing
Estimator-correlator receiver, 3
Estimator-subtractor receiver, 63, 118

Factorization
spectral, 17, 29
covariance, 18, 19
Faddeeva algorithm, 30, 39, 170–171
Filtering. *See also* Smoothing
delay, 4, 121, 144–155
maximum *a posteriori*, 6, 127, 156–169
nonlinear, 3, 6, 156–169
realizable, 121, 122, 127–128, 163
state variable processes, 2, 4, 6, 32, 67, 122–155, 156–169
stationary processes, 2, 154
unrealizable. *See* Smoothing
First-order processes. *See* Butterworth processes; Wiener process
Fokker-Planck equation, 167
Fourier coefficient, 23
Fredholm determinant, 5, 31–34
Fredholm integral equations. *See* Integral equations
Frequency modulation, 161

Gaussian statistics, 10, 122, 127, 156–157, 161
Global optimality, 96, 99–100, 120
Gradient matrices, 157

Hamiltonian. *See* Minimum principle
Hard constraints. *See* Constraints for signal design
Hermitian matrices, 174, 177

Impulse response for filters and smoothers, 123
Independent increment processes, 9, 130
Integral equations
first kind, 59, 77, 79
homogeneous, 3–5, 23–55
signal design, 81, 93–95
solution properties, 24
inhomogeneous, 3–5, 56–80
smoother solution, 125, 127, 155
solution methods, 63–72
Invariant imbedding, 164–168

Kalman Bucy filtering. *See* Filtering, state variable processes
Karhunen-Loève expansion, 2, 23. *See also* Integral equations, homogeneous

Kellog's method, 31

Lag filtering. *See* Filtering, delay
λ_E-λ_B (energy multiplier-bandwidth multiplier) plane, 100–105, 110
Laplace methods, 30
Likelihood function, 3

Markov processes, 13, 34
M-Ary signals, 120
Matched filters, 77, 102, 107–108
  optimally designed, 108–109
Matrix exponentials, 14, 36, 170–171. *See also* Transition matrices
Means
  Doppler shifts, 180, 183
  process, 123
Minimum principle, 84–96, 118
  Hamiltonian for signal design, 88, 114
  necessary conditions, 88, 114
  stochastic, 118
Modulation operation, 157
  observation equations, 9
Multimodal solutions. *See* Global optimality
Multiple-order solutions. *See also* Minimum principle
  eigenvalues, 29, 41

Nesting, 171
Nonstationary first-order processes, 38–39
Notation, 7
  complex for narrowband processes, 172–173
Number of computations, 80
Numerical methods, 30, 43, 68, 74, 79–80
  stability, 69–70, 100–102

Optimal control theory. *See* Minimum principle
Order function, 32
Orthonormal expansion. *See* Karhunen-Loève expansion
Orthogonality principle, 131

Parameter estimation bounds, 54, 56
  Bhattacharyya, 54
  Cramer-Rao, 54
Pathological solutions, signal design, 94
Performance, improvement by delay, 139, 145. *See also* Detection, performance; Covariance matrix, error
Pontryagin's minimum principle. *See* Minimum principle
Pre-emphasis, 140
Process generation, 8–18
Process representation, impulse response-covariance method, 16. *See also* State variables, description
Prolate spheroidal waveforms, 44

Quasi-linearization, 163

Rank, signal design determinant, 99, 101
Receivers
  detectors in colored noise, 57–58
  detectors in clutter, 57
Reduction, dimension of signal design equations, 92–93
Reverberation, Doppler spread, 60, 82, 96. *See also* Detection in clutter
Ricatti equation. *See* Covariance matrix; Filtering

Scattering function, 120
Second-order processes, 16, 39–43, 75–77, 109–114, 140–144, 180–182
Semidefinite covariances, 26
Signal bandwidth. *See* Bandwidth; Constraints for signal design
Signal design, 5, 55, 81–120
  additive channels, 97–115, 120
  bandpass channels, 119
  Doppler spread channels, 115–120
Singularity functions
  integral equations of first kind, 59, 77, 79
  turn on–turn off, 87
  white noise covariances, 8
Smoothing
  linear, 4, 56, 63, 122–155
  nonlinear, 4, 156–169
Solution methods
  inhomogeneous integral and linear smoothing equations, 63–70
  nonlinear smoothing, 162–163
Stability of solutions, 69–70, 131, 132, 149, 155

State variables
 complex. *See* Bandpass process representation
 covariances, 10–18
 descriptions, 8–11
  initial state, 9
  observation equation, 9
  state equation, 9
  state vector, 9
 random processes, 8–18
 solution
  homogeneous integral equation, 24–31
  inhomogeneous integral equation, 62–70
Stationary processes
 constant parameter systems, 14
 narrow-band processes, 172
 rational kernels, 26
 spatial covariances, 14
Structual approaches, 10, 122
Suboptimal solutions, signal design, 105
Sufficient statistics. *See* Receivers
Superposition method, 64, 163
Symmetry, optimal signals, 105, 107

Tupped delay lines, 149
Time-bandwidth products, 34, 39, 41, 119, 171
Time varying coefficients, optimal smoother, 127, 134
Tracking, signal design for, 119. *See also* Smoothing
Transition matrices
 definition, 12
 matrix exponentials, 14, 30, 101, 170–171

Unstructured Gaussian approaches, 10, 122

White noise, 8, 59, 161
Whitening filters, 128–130
Wiener filtering. *See* Filtering, stationary processes
Wiener-Hopf equation, 3, 56, 121, 123, 155
Wiener process, 15, 35–36, 70–72, 134–135